Author:

Ruth M. Young, M.S. Ed.,
Elementary Science
Consultant

Illustrator:

Blanca Apodaca La Bounty

Editors:

Evan D. Forbes, M.S. Ed.
Walter Kelly, M.A.

Senior Editor:

Sharon Coan, M.S. Ed.

Art Direction:

Elayne Roberts

Product Manager:

Phil Garcia

Imaging:

Alfred Lau

Research:

Bobbie Johnson

Cover Photo Credit:

Images © PhotoDisc, Inc., 1994

Publishers:

Rachelle Cracchiolo, M.S. Ed.
Mary Dupuy Smith, M.S. Ed.

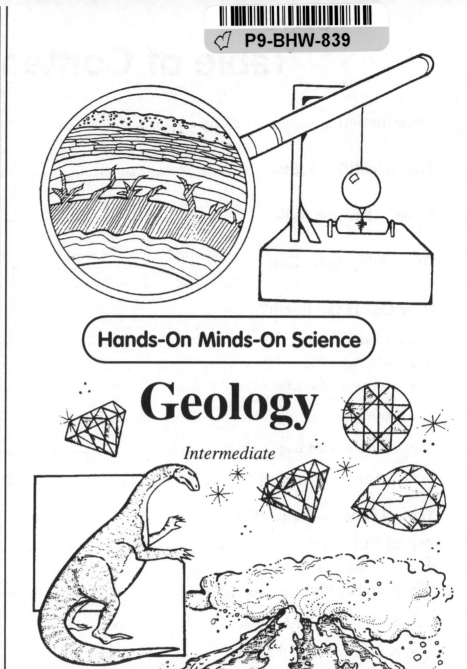

Hands-On Minds-On Science

Geology

Intermediate

Teacher Created Materials, Inc.

6421 Industry Way
Westminster, CA 92683
www.teachercreated.com

©1994 Teacher Created Materials, Inc.

Reprinted, 1999

Made in U.S.A.

ISBN-1-55734-641-0

Inside the earth
volcano (2 day)
Changes - 50 million (3-4)
Earth crust (2-3)
years
Earthquakes
2-3

Table of Contents

Table of Contents *(cont.)*

Introduction

What Is Science?

What is science to children? Is it something that they know is a part of their world? Is it a textbook in the classroom? Is it a tadpole changing into a frog? Is it a sprouting seed, a rainy day, a boiling pot, a turning wheel, a pretty rock, or a moonlit sky? Is science fun and filled with wonder and meaning? What does science mean to children?

Science offers you and your eager students opportunities to explore the world around you and to make connections between the things you experience. The world becomes your classroom, and you, the teacher, a guide.

Science can, and should, fill children with wonder. It should cause them to be filled with questions and the desire to discover the answers to their questions. And, once they have discovered the answers, they should be actively seeking new questions to answer!

The books in this series give you and your students the opportunity to learn from the whole of your experience—sights, sounds, smells, tastes, and touches, as well as what you read, write about, and do. This whole science approach allows you to experience and understand your world as you explore science concepts and skills together.

What Is Geology?

Geology literally translated means the study of the earth. This book covers only a few of the many areas of geology. These include the age and evolution of earth, the cause and location of earthquakes, the movement of the continents over millions of years, and the composition of the earth's interior. Like a search for buried treasure, the study of geology can unearth gems and nuggets of knowledge and insight. We cannot help turning them over in our hands, marveling at their heft, gleam, and luminosity. To see the jutting mountains of our earth, the smoke snaking from a volcano, to sense in our bodies the heave and jolt of an earthquake is to be captured by a desire to learn more about our living planet.

The history of our earth—and much remains to be learned—has revealed amazing things, from the existence of dinosaurs to the causes of volcanoes, earthquakes, and ice ages. Every bit we learn whets our appetite to know more of the past while sharpening our impulse to predict the future. Geology is indeed a proper study for all humankind.

The Scientific Method

The "scientific method" is one of several creative and systematic processes for proving or disproving a given question, following an observation. When the "scientific method" is used in the classroom, a basic set of guiding principles and procedures is followed in order to answer a question. However, real world science is often not as rigid as the "scientific method" would have us believe.

This systematic method of problem solving will be described in the paragraphs that follow.

1 Make an OBSERVATION.

The teacher presents a situation, gives a demonstration, or reads background material that interests students and prompts them to ask questions. Or students can make observations and generate questions on their own as they study a topic.

Example: Observing a large crack in the earth's crust.

2 Select a QUESTION to investigate.

In order for students to select a question for a scientific investigation, they will have to consider the materials they have or can get, as well as the resources (books, magazines, people, etc.) actually available to them. You can help them make an inventory of their materials and resources, either individually or as a group.

Tell students that in order to successfully investigate the questions they have selected, they must be very clear about what they are asking. Discuss effective questions with your students. Depending upon their level, simplify the question or make it more specific.

Example: Does the crust move?

3 Make a PREDICTION (Hypothesis).

Explain to students that a hypothesis is a good guess about what the answer to a question will probably be. But they do not want to make just any arbitrary guess. Encourage students to predict what they think will happen and why.

In order to formulate a hypothesis, students may have to gather more information through research.

Have students practice making hypotheses with questions you give them. Tell them to pretend they have already done their research. You want them to write each hypothesis so it follows these rules:

1. It is to the point.
2. It tells what will happen, based on what the question asks.
3. It follows the subject/verb relationship of the question.

Example: I think as the earth moves inside, it causes the crust to move on the outside.

The Scientific Method *(cont.)*

4 Develop a PROCEDURE to test the hypothesis.

The first thing students must do in developing a procedure (the test plan) is to determine the materials they will need.

They must state exactly what needs to be done in step-by-step order. If they do not place their directions in the right order, or if they leave out a step, it becomes difficult for someone else to follow their directions. A scientist never knows when other scientists will want to try the same experiment to see if they end up with the same results!

Example: With the use of peanut butter and jelly sandwiches, you will simulate the movement and layers of the earth's crust.

5 Record the RESULTS of the investigation in written and picture form.

The results (data collected) of a scientific investigation are usually expressed two ways—in written form and in picture form. Both are summary statements. The written form reports the results with words. The picture form (often a chart or graph) reports the results so the information can be understood at a glance.

Example: The result of this investigation can be recorded on a data-capture sheet provided (page 57).

6 State a CONCLUSION that tells what the results of the investigation mean.

The conclusion is a statement which tells the outcome of the investigation. It is drawn after the student has studied the results of the experiment, and it interprets the results in relation to the stated hypothesis. A conclusion statement may read something like either of the following: "The results show that the hypothesis is supported," or "The results show that the hypothesis is not supported." Then restate the hypothesis if it was supported or revise it if it was not supported.

Example: The hypothesis that stated "as the earth moves inside, it causes the crust to move on the outside is supported (or not supported).

7 Record QUESTIONS, OBSERVATIONS, and SUGGESTIONS for future investigations.

Students should be encouraged to reflect on the investigations that they complete. These reflections, like those of professional scientists, may produce questions that will lead to further investigations.

Example: What happens when the earth's crust moves?

Science-Process Skills

Even the youngest students blossom in their ability to make sense out of their world and succeed in scientific investigations when they learn and use the science-process skills. These are the tools that help children think and act like professional scientists.

The first five process skills on the list below are the ones that should be emphasized with young children, but all of the skills will be utilized by anyone who is involved in scientific study.

Observing

It is through the process of observation that all information is acquired. That makes this skill the most fundamental of all the process skills. Children have been making observations all their lives, but they need to be made aware of how they can use their senses and prior knowledge to gain as much information as possible from each experience. Teachers can develop this skill in children by asking questions and making statements that encourage precise observations.

Communicating

Humans have developed the ability to use language and symbols which allow them to communicate not only in the "here and now" but also over time and space as well. The accumulation of knowledge in science, as in other fields, is due to this process skill. Even young children should be able to understand the importance of researching others' communications about science and the importance of communicating their own findings in ways that are understandable and useful to others. The geology journal and the data-capture sheets used in this book are two ways to develop this skill.

Comparing

Once observation skills are heightened, students should begin to notice the relationships between things that they are observing. *Comparing* means noticing similarities and differences. By asking how things are alike and different or which is smaller or larger, teachers will encourage children to develop their comparison skills.

Ordering

Other relationships that students should be encouraged to observe are the linear patterns of seriation (order along a continuum: e.g., rough to smooth, large to small, bright to dim, few to many) and sequence (order along a time line or cycle). By ranking graphs, time lines, cyclical and sequence drawings, and by putting many objects in order by a variety of properties, students will grow in their abilities to make precise observations about the order of nature.

Categorizing

When students group or classify objects or events according to logical rationale, they are using the process skill of categorizing. Students begin to use this skill when they group by a single property such as color. As they develop this skill, they will be attending to multiple properties in order to make categorizations; the animal classification system, for example, is one system students can categorize.

Science-Process Skills *(cont.)*

Relating

Relating, which is one of the higher-level process skills, requires student scientists to notice how objects and phenomena interact with one another and the change caused by these interactions. An obvious example of this is the study of chemical reactions.

Inferring

Not all phenomena are directly observable, because they are out of humankind's reach in terms of time, scale, and space. Some scientific knowledge must be logically inferred based on the data that is available. Much of the work of paleontologists, astronomers, and those studying the structure of matter is done by inference.

Applying

Even very young, budding scientists should begin to understand that people have used scientific knowledge in practical ways to change and improve the way we live. It is at this application level that science becomes meaningful for many students.

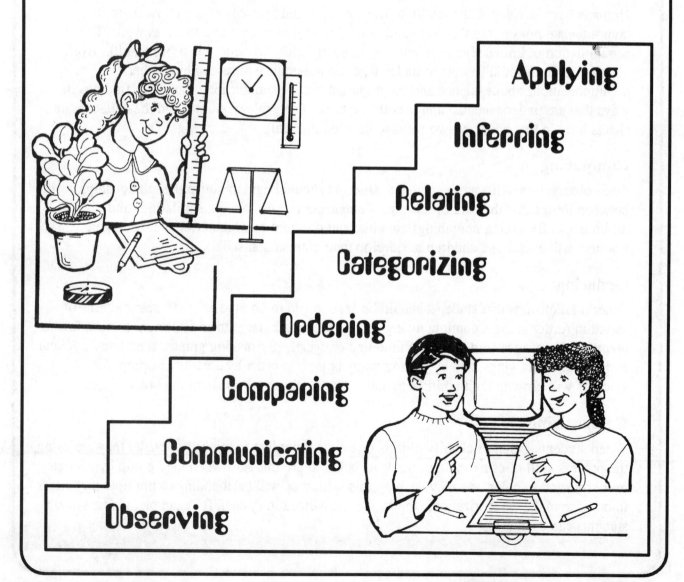

Applying

Inferring

Relating

Categorizing

Ordering

Comparing

Communicating

Observing

8

Organizing Your Unit

Designing a Science Lesson

In addition to the lessons presented in this unit, you will want to add lessons of your own, lessons that reflect the unique environment in which you live, as well as the interests of your students. When designing new lessons or revising old ones, try to include the following elements in your planning:

Question

Pose a question to your students that will guide them in the direction of the experience you wish to perform. Encourage all answers, but you want to lead the students towards the experiment you are going to be doing. Remember, there must be an observation before there can be a question. (Refer to The Scientific Method, pages 5-6.)

Setting the Stage

Prepare your students for the lesson. Brainstorm to find out what students already know. Have children review books to discover what is already known about the subject. Invite them to share what they have learned.

Materials Needed for Each Group or Individual

List the materials each group or individual will need for the investigation. Include a data-capture sheet when appropriate.

Procedure

Make sure students know the steps to take to complete the activity. Whenever possible, ask them to determine the procedure. Make use of assigned roles in group work. Create (or have your students create) a data-capture sheet. Ask yourself, "How will my students record and report what they have discovered? Will they tally, measure, draw, or make a checklist? Will they make a graph? Will they need to preserve specimens?" Let students record results orally, using a video or audio tape recorder. For written recording, encourage students to use a variety of paper supplies such as poster board or index cards. It is also important for students to keep a journal of their investigation activities. Journals can be made of lined and unlined paper. Students can design their own covers. The pages can be stapled or be put together with brads or spiral binding.

Extensions

Continue the success of the lesson. Consider which related skills or information you can tie into the lesson, like math, language arts skills, or something being learned in social studies. Make curriculum connections frequently and involve the students in making these connections. Extend the activity, whenever possible, to home investigations.

Closure

Encourage students to think about what they have learned and how the information connects to their own lives. Prepare geology journals using directions on page 81. Provide an ample supply of blank and lined pages for students to use as they complete the Closure activities. Allow time for students to record their thoughts and pictures in their journals.

Organizing Your Unit *(cont.)*

Structuring Student Groups for Scientific Investigations

Using cooperative learning strategies in conjunction with hands-on and discovery learning methods will benefit all the students taking part in the investigation.

Cooperative Learning Strategies

1. In cooperative learning, all group members need to work together to accomplish the task.
2. Cooperative learning groups should be heterogeneous.
3. Cooperative learning activities need to be designed so that each student contributes to the group and individual group members can be assessed on their performance.
4. Cooperative learning teams need to know the social as well as the academic objectives of a lesson.

Cooperative Learning Groups

Groups can be determined many ways for the scientific investigations in your class. Here is one way of forming groups that has proven to be successful in intermediate classrooms.

- **The Team Leader**—scientist in charge of reading directions and setting up equipment.
- **The Geologist**—scientist in charge of carrying out directions (can be more than one student).
- **The Stenographer**—scientist in charge of recording all of the information.
- **The Transcriber**—scientist who translates notes and communicates findings.

If the groups remain the same for more than one investigation, require each group to vary the people chosen for each job. All group members should get a chance to try each job at least once.

Using Centers for Scientific Investigations

Set up stations for each investigation. To accommodate several groups at a time, stations may be duplicated for the same investigation. Each station should contain directions for the activity, all necessary materials (or a list of materials for investigators to gather), a list of words (a word bank) which students may need for writing and speaking about the experience, and any data-capture sheets or needed materials for recording and reporting data and findings.

Model and demonstrate each of the activities for the whole group. Have directions at each station. During the modeling session, have a student read the directions aloud while the teacher carries out the activity. When all students understand what they must do, let small groups conduct the investigations at the centers. You may wish to have a few groups working at the centers while others are occupied with other activities. In this case, you will want to set up a rotation schedule so all groups have a chance to work at the centers.

Assign each team to a station, and after they complete the task described, help them rotate in a clockwise order to the other stations. If some groups finish earlier than others, be prepared with another unit-related activity to keep students focused on main concepts. After all rotations have been made by all groups, come together as a class to discuss what was learned.

Just the Facts

How old are you? What exciting events have happened during your lifetime? You have changed much since the day you were born. Have you ever wondered if the earth changed as it aged also? Have you ever been curious about the age of the earth? Answering these questions about the earth is like solving a mystery. Scientists collect evidence from the rocks all over the world, including those on the ocean floor. They put all of this evidence together to determine the earth's age and describe the changes it went through from its earliest stages to present day.

Most scientists think the earth is at least 4.5 billion years old. The oldest rocks found on earth are 4.3 billion years old. Rocks formed prior to that time would have been melted during the early formation of earth. The age of rocks is computed from the number of radioactive isotopes they contain.

It is believed the solar system developed from a huge spiraling nebula (cloud of gas and dust-sized pieces of rock and metal). The sun formed in the center of this nebula which gradually flattened as it spun and developed into whirlpools of gas that collected together as the planets. Our planet began at the same time as all the other planets in the solar system.

Many changes gradually took place as the earth evolved over billions of years. The shapes and locations of continents have changed and continue to change as the crust of the earth shifts. This has had a profound effect on the plant and animal life, as well as the geology of the earth's surface. Mountain ranges formed, islands were born and destroyed, land bridges between continents were pulled apart, and land masses were slowly relocated and reshaped.

When maps that showed the shapes of all the continents first became available, some scientists suggested that the pieces could fit together like puzzle pieces, particularly South America and Africa. Maps based upon scientific evidence of fossil remains and geological changes show very different early locations for the continents. Approximately 200 million years ago the continents were joined together as one. That early land mass has been named Pangaea. Later, Pangaea divided into a northern continent, Laurasia, and a southern continent, Gondwanaland.

We are all familiar with the extinction of the dinosaurs about 65 million years ago. Theories are that this extinction may have been caused by a meteor striking earth, sending debris into the atmosphere, drastically changing the climate and destroying many species of plants and animals. Not every plant or animal died during this period, nor was this the only extinction event in earth's long history.

The activities in this section will unfold the history of these changes which have happened on earth. They include making a time line of the earth's history and analyzing maps of the earth's past.

Making a Time Line

Question

What is a time line?

Setting the Stage

- Discuss with students the need to understand order and sequence. Ask students whether our lives would be different if some events were reversed. Ask for specific examples.
- Ask students if events that come before sometimes cause events that follow. Ask for actual examples in history, science, and real-life behavior. Use the board to list brainstormed examples.
- Distribute a copy of the data-capture sheets (pages 13-14) to each student several days prior to doing this lesson. Be sure they have this form completed for the activity.

Materials Needed for Each Individual

- 1 yd (1 m) length of 2" (5 cm) wide adding machine tape
- metric or standard ruler
- data-capture sheets (pages 13-14), completed

Note to the teacher: Students will make a time line of their lives to introduce them to this technique of describing the sequence and length of historical events.

Teacher Preparation for Activity

Cut two strips of adding machine tape, each 1 yd (1 m) long. Mark strips as shown:

1 yd (1 m) = 1 year:

.... 1 years

1 yd (1 m) = 10 years:

| 1 | 2 | 3 | 4 | 5 | 6 | 7 | 8 | 9 | 10 years |

Procedure

1. Tape the time line strip for one year across the board.
2. Have one student read one year of his/her life from their data-capture sheets while you record the events along the strip on the board in the order in which they happened.
3. Tape the time line strip for ten years across the board. Show how the one year of information would fit on the ten year strip.
4. Distribute blank strips and rulers to students. Show them how to divide the strips into sections that look like the strip on the board.
5. After students label years from one to ten along the division marks, have them record the information from their data-capture sheets (pages 13-14) on the time line strip. (Students older than ten may need to add paper.)

Extension

Have students bring in photographs which will illustrate the events of their life and show how they changed through the years. Display the students' time lines on the bulletin board.

Closure

Divide students into groups of three or four and have them share their time lines.

Making a Time Line *(cont.)*

Name:_____ Birthdate: ____ / ____ / ____

month date year

Fill in the summary of your life on the form below for our next science class. You will use this information to create your own time line. Important events may include such things as when you first had teeth, walked, spoke, took special trips, or had health problems such as broken bones.

Age in Years	Where You Lived	Important Events (At least three per year if possible.)	Date
0-1	_____	_____	___ / ___ / ___
		_____	___ / ___ / ___
		_____	___ / ___ / ___
		_____	___ / ___ / ___
1-2	_____	_____	___ / ___ / ___
		_____	___ / ___ / ___
		_____	___ / ___ / ___
		_____	___ / ___ / ___
2-3	_____	_____	___ / ___ / ___
		_____	___ / ___ / ___
		_____	___ / ___ / ___
		_____	___ / ___ / ___
3-4	_____	_____	___ / ___ / ___
		_____	___ / ___ / ___
		_____	___ / ___ / ___
		_____	___ / ___ / ___
4-5	_____	_____	___ / ___ / ___
		_____	___ / ___ / ___
		_____	___ / ___ / ___
		_____	___ / ___ / ___

Making a Time Line *(cont.)*

Name:_____   Birthdate: ____ / ____ / ____

month   date   year

Age in Years	Where You Lived	Important Events (At least three per year if possible.)	Date
5-6	_____	_____	___ / ___ / ___
		_____	___ / ___ / ___
		_____	___ / ___ / ___
		_____	___ / ___ / ___
6-7	_____	_____	___ / ___ / ___
		_____	___ / ___ / ___
		_____	___ / ___ / ___
		_____	___ / ___ / ___
7-8	_____	_____	___ / ___ / ___
		_____	___ / ___ / ___
		_____	___ / ___ / ___
		_____	___ / ___ / ___
8-9	_____	_____	___ / ___ / ___
		_____	___ / ___ / ___
		_____	___ / ___ / ___
		_____	___ / ___ / ___
9-10	_____	_____	___ / ___ / ___
		_____	___ / ___ / ___
		_____	___ / ___ / ___
		_____	___ / ___ / ___
10-11	_____	_____	___ / ___ / ___
		_____	___ / ___ / ___
		_____	___ / ___ / ___
		_____	___ / ___ / ___

Earth's Time Line

Question

What does a time line for the earth look like?

Setting the Stage

- Remind students of how they made a time line for their lives.
- Provide them with the background information on the age of the earth shown in the Just the Facts section (page 11).
- Ask students how long the time line would need to be for the 4.5 billion year old earth if the scale is 1 yd (1 m) = 100,000,000 years. (The time line would need to be 45 yds (45 m) long.)
- Explain to students that they will work together to make this time line, put it together, and then take a "walk through time."
- Project for students a transparency copy of Earth's Time Line Information (pages 17-18) and explain that this is what the groups will use to make their time lines.
- Tell students that all measurements should be carefully made. Describe for students a system to be used for marking each section to show the division of eras, periods, and millions of years.
- Show students how information from "Life Forms" column should be written on each section.

Materials Needed For Each Group

- pencils
- felt markers
- 3" (8 cm) adding machine tape sections as shown

- Earth's Time Line Information (pages 17-18), one each per student
- data-capture sheet (page 19), one per student

Group #	Number of Students	Section	Length of Tape or Stick	Measuring Device(s)
1	6	Cenozoic Era	1 yd (1 m)	metric or standard tape measure
2	4	Mesozoic Era	2 yds (2 m)	metric or standard tape measure
3	5	240-435 mya* Paleozoic Era	2 yds (2 m)	metric or standard tape measure
4	5	435-900 mya Paleozoic Era	5 yds (5 m)	metric or standard tape measure
5	3	900-4,000 mya Paleozoic Era	31 yds (31 m)	meter or yard stick
6	3	4,000-4,500 mya Paleozoic/ Pre-Cambrian Eras	6 yds (6 m)	meter or yard stick

*mya=million(s) of years ago

Note to the teacher: Students will make a time line of the earth's 4.5 billion year history.

Earth's Time Line *(cont.)*

Procedure

1. Distribute to students a copy of the Earth's Time Line Information (pages 17-18) which has been marked to show the section each group will follow when making their time line.

2. Explain that students are to work together to measure their adding machine tape according to the information listed in the "Length from last mark" column (pages 17-18).

3. Monitor students as they work and offer assistance when necessary.

4. Have students complete their data-capture sheets.

Extensions

• Tape the ends of each group's time line to the next in the order of years. Then stretch out the entire tape of your class "walk through time." Begin the walk 4.5 billion years ago, letting a spokesperson from each group read the information about their period during the walk.

• Have students research a specific period of time and then report to the class.

Closure

In their geology journals, have students write a paragraph about living during the early part of the earth's history.

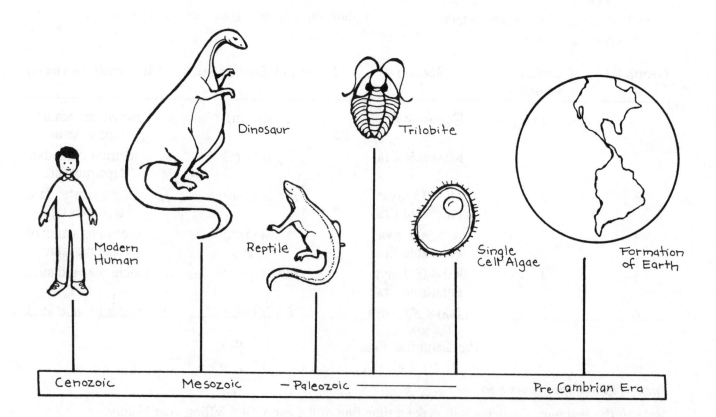

Earth's Time Line *(cont.)*

Earth's Time Line Information

Scale: 1 yd (1 m) = 100 million years (.4" (1 cm) = 1 million years)

	Period	Millions of Years Ago	Life Forms	(Length from last mark)
C E N O Z O I C **E R A**	Quaternary	1-1.5	Homo Erectus.	.4" (1 cm)
		2	Modern humans began to develop; mammoths.	.4" (1 cm)
		4	Australopithecus, closest primate ancestor to man appears in Africa.	.8" (2 cm)
	Tertiary	10	Ramapithecus, oldest known primate with human-like traits, evolved in India and Africa.	2.4" (6 cm)
		40	Monkeys and apes evolve.	12" (30 cm)
		70	Prosimians, earliest primates develop.	12" (30 cm)

Extinction at end of Cretaceous Period, possibly by meteorite impact, brings many marine species and dinosaur era to end.

	Period	Millions of Years Ago	Life Forms	(Length from last mark)
M E S O Z O I C **E R A**	Cretaceous	138	Triceratops, Tyrannosaurus Rex, flowering plants.	27" (69 cm)
	Jurassic	205	Largest dinosaurs Stegosaurus, Aptosaurus, flying reptiles, birds.	27" (69 cm)

Extinction at end of Triassic Period: up to 75% of marine invertebrate species and some land dwellers vanish, perhaps in an impact-related catastrophe. Two recently evolved groups, dinosaurs and mammals, survive.

	Period	Millions of Years Ago	Life Forms	(Length from last mark)
P A L E O Z O I C **E R A**	Triassic	240	Dinosaurs first appeared, earliest mammals.	14" (35 cm)

Greatest mass extinction of all time, nearly all Permian Period species perish.

Period	Millions of Years Ago	Life Forms	(Length from last mark)
Permian	290	Eryops, Dimetrodon, Thecodont, ancestor to dinosaurs.	20" (50 cm)
Carboniferous	330	Amphibians, reptiles, land plants, trees, insects.	16" (40 cm)

Earth's Time Line *(cont.)*

Earth's Time Line Information

Scale: 1 yd (1 m) = 100 million years (.4" (1 cm) = 1 million years)

	Period	Millions of Years Ago	Life Forms	(Length from last mark)
			Extinction of most of world's fish, and perhaps 70% of its invertebrate species perish in late Devonian Period.	
P A L E O Z O I C	Devonian	410	Small amphibians venture onto land, ocean plants.	38" (97 cm)
	Silurian	435	Armored fish, first vertebrates.	10" (25 cm)
			Supercontinent Gondwanaland drifts over South Pole in Ordovician, triggering period of prolonged glaciation. Early fish survive, but marine invertebrates and primitive reef builders are hard hit.	
	Ordovician	500	Horseshoe crabs, sharks.	24" (60 cm)
	Cambrian	570	Trilobites, shell bearing, multi-celled invertebrates.	28" (70 cm)
		700	Worms and jelly fish.	1.3 yds (1.3 m)
P R E- C A M B R I A N		900	First oxygen-breathing animals in ocean.	2 yds (2 m)
		3,500	Life first appeared in ocean: single celled algae and bacteria.	26 yds (26 m)
		4,000	Formation of primordial sea, no life forms.	5 yds (5 m)
		4,500	Earth formation from gaseous material spun off the sun.	5 yds (5 m)

Earth's Time Line *(cont.)*

Use Earth's Time Line Information chart (pages 17-18) to fill in the information needed.

Dinosaurs became extinct about 65 million years ago, at the end of the Cretaceous Period. Complete the chart by describing the extinctions which took place before the dinosaur extinction. That extinction is shown as an example.

When Extinction Occurred   **What Happened**

End of Cretaceous Period   Dinosaur era ends and many marine species die, possibly due to meteorite hitting earth.

_____   _____

_____   _____

_____   _____

_____   _____

_____   _____

_____   _____

_____   _____

The scale used on the Earth's Time Line is _____

Use the *Earth's Time Line Information* chart to complete the following:

Modern humans have lived on earth about _____ years.

Dinosaurs lived on earth about _____ years ago.

The earliest mammals first appeared on earth about _____years ago.

Time from beginning of earth until first life appeared was about _____years.

How long had mammals been on earth before modern humans developed? _____

Changes in the Earth's Crust

Question

Has the earth's surface changed during its 4.5 billion year history?

Setting the Stage

- Ask students if they have ever looked carefully at the shapes of some of the continents.
- Project for students a transparency of the continents and point out the shapes of North America, South America, Africa, Europe, and Asia.
- Explain to students they will use puzzle pieces of these continents to see if they can discover what was noticed by some people who looked at maps of these continents when they were first made.

Materials Needed for Each Group

- copy of Continent Puzzle (page 21), one per student
- scissors

    Note to the teacher: Using the continents of North America, South America, Africa, Europe, and Asia as puzzle pieces, students will discover that perhaps they could have fit together at one time. Maps showing the location of the continents from 200 million years ago to 50 million years in the future will be examined later and analyzed.

Procedure

1. Have students carefully cut out the Continent Puzzle as follows:
   - North and South America as one piece
   - Africa, Europe, and Asia as another piece
2. Provide your students with the time necessary to play with these puzzle pieces to see if they can fit them together.

Extension

Share with students the information from Just the Facts (page 11) about the evidence scientists have found to prove that the continents were once grouped as one.

Closure

Have students share what they discovered with the entire class. (North and South America appear to fit with Africa and Europe.)

Changes in the Earth's Crust *(cont.)*

Continent Puzzle

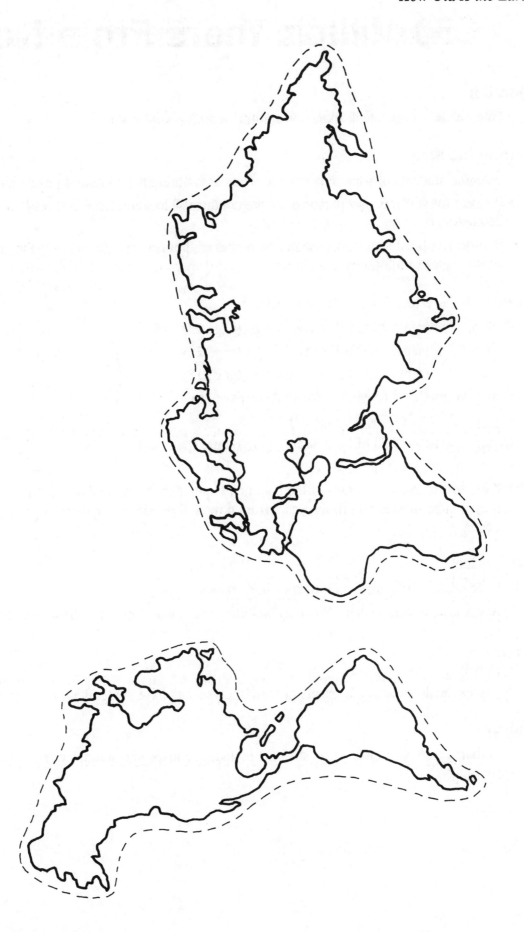

50 Million Years From Now

Question

What will the earth look like in 50 million years?

Setting the Stage

- Remind students of what they learned while assembling the continent puzzle pieces.
- Project for students transparencies of the continents' location from 200 million years ago to the present.
- Discuss with students the changes which appeared during this period in the location of the continents and life forms.

Materials Needed for Each Individual

- map: Continents 200 Million Years Ago (page 23)
- map: Continents 180 Million Years Ago (page 24)
- map: Continents 135 Million Years Ago (page 25)
- map: Continents 65 Million Years Ago (page 26)
- map: Continents Today (page 27)
- map: Continents 50 Million Years in the Future (page 28)
- data-capture sheet (page 29)

  Note to the teacher: Maps of the continents' positions today and a scientific projection of where they will be located 50 million years from now will be used to analyze the changes that will occur.

Procedure

1. Have students complete their data-capture sheets.
2. Assist students as needed. You may wish to show a world map for them to compare to the drawings.

Extension

Discuss with students the results of their analysis and have students write additional observations.

Closure

In their geology journals, have students include the maps and summary sheets they have completed.

50 Million Years From Now *(cont.)*

Continents 200 Million Years Ago

Mesozoic Era
Triassic Period

Cone-bearing trees were plentiful. Many fish resembled those of modern times. Insects were plentiful. The first turtles, crocodiles, dinosaurs, and mammals appeared.

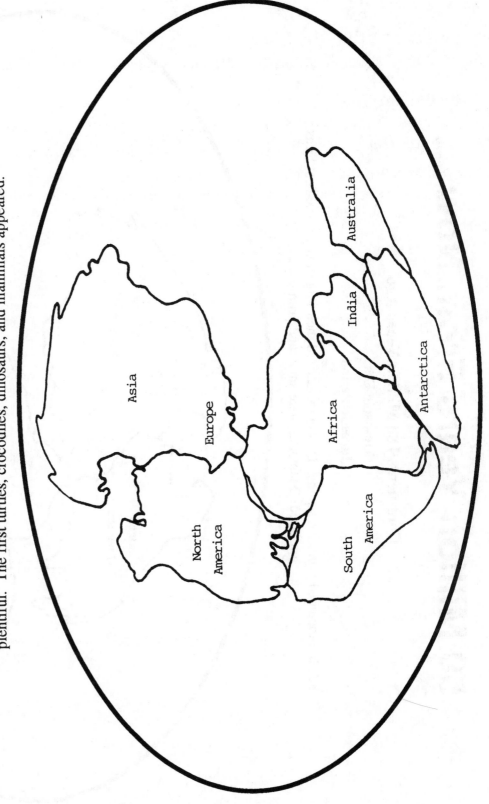

50 Million Years From Now *(cont.)*

Continents 180 Million Years Ago

Mesozoic Era

Jurassic Period

Sea life included primitive squids. Dinosaurs reached their largest size. The first birds appeared. Mammals were small and primitive.

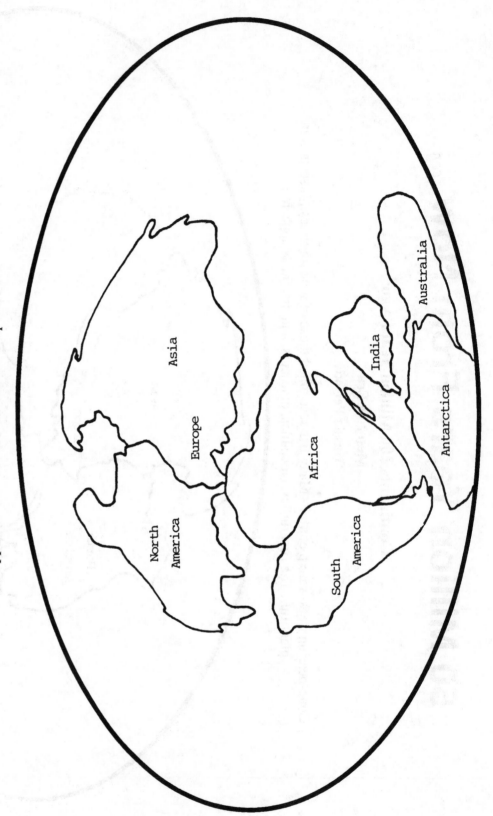

50 Million Years From Now *(cont.)*

Continents 135 Million Years Ago

Mesozoic Era

Cretaceous Period

Flowering plants appeared. Invertebrates, fish, and amphibians were plentiful. Dinosaurs with horns and armor became common. Dinosaurs died out at the end of this period.

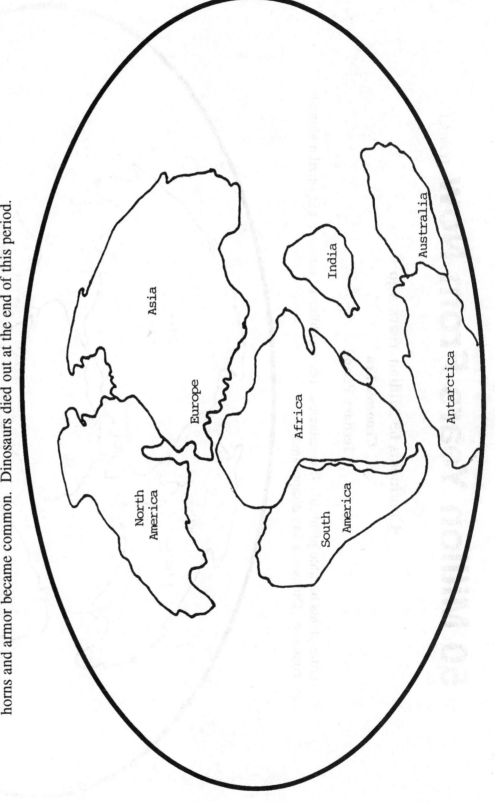

50 Million Years From Now *(cont.)*

Continents 65 Million Years Ago
Cenozoic Era
Tertiary Period

Flowering plants became plentiful. Invertebrates, fish, amphibians, reptiles, and small mammals were common. Dinosaurs had disappeared.

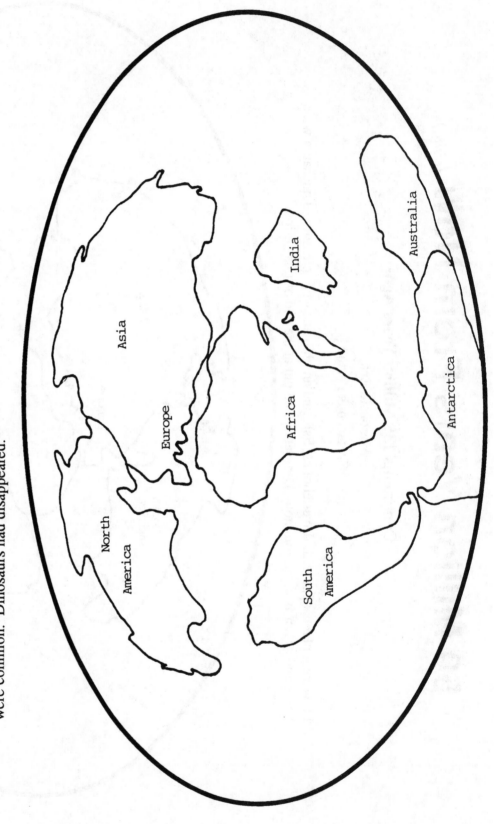

50 Million Years From Now *(cont.)*

Continents Today

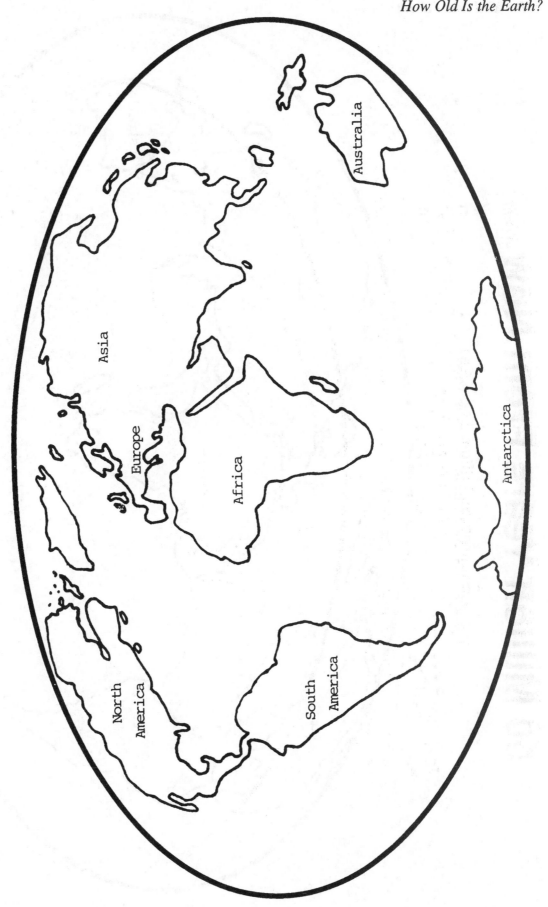

50 Million Years From Now *(cont.)*

Continents 50 Million Years in the Future

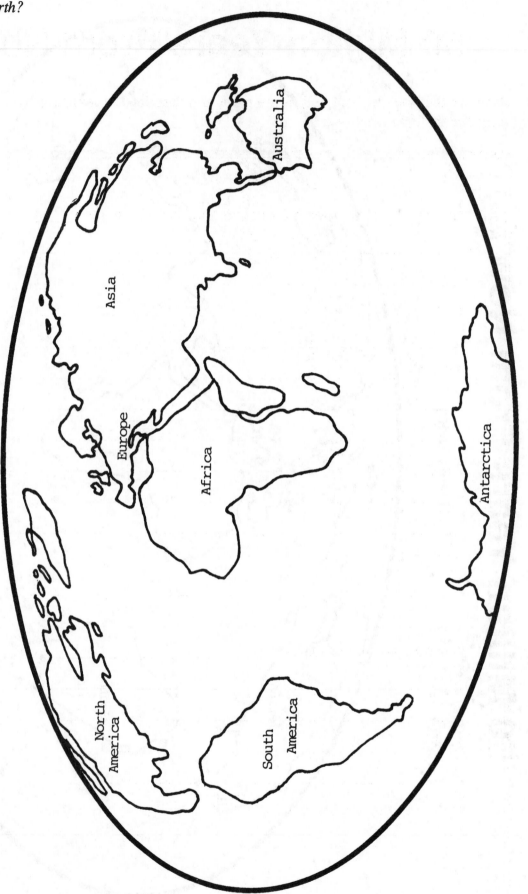

50 Million Years From Now *(cont.)*

Compare the past, present, and future maps and describe the changes which are predicted to occur for the following areas:

North America, including California and Baja, CA areas _____

_____

_____

South America _____

_____

Africa _____

_____

_____

Europe and Saudi Arabia _____

_____

_____

Asia (including India) _____

_____

_____

Australia and New Guinea _____

_____

_____

Japan _____

_____

How do you think this will change life on earth? _____

_____

_____

_____

Just the Facts

You have probably heard of the water cycle, but do you know that rocks are recycled too? What does the earth look like inside?

How do we know what is inside the earth? Can we drill into the center of the earth?

All the rocks which are on earth have been here since earth first began about 4.5 billion years ago. The only new rocks are occasional meteors which are pulled toward earth by gravity as they come near our planet. (Of course, we also have a few rocks which were brought back by astronauts who first landed on the moon.) Rocks are constantly being formed as magma from beneath the earth's crust, which then forces its way to the surface in volcanoes. Rock is recycled through erosion forces of water, wind, and ice. Rock is also melted when it goes deep enough to make contact with the mantle of the earth.

Some of the new words you will find in this section are defined below:

Crust: The hard layer of rock which covers the outside of the earth. This layer is about 5–20 miles (8–32 km) thick and lies just above the mantle.

Mantle: The layer just below the crust is about 1,800 miles (2,880 km) thick and consists of semisolid rock which is approximately 1,600° F (871° C). This extremely hot taffy-like layer moves slowly in convection currents between the crust and outer core.

Magma: The semisolid material in the mantle which sometimes pours through cracks in the crust and is then called lava. When it cools and hardens into rock, it is referred to as igneous rock.

Outer Core: Just below the mantle, this 1,400 miles (2,240 km) layer is made of liquid iron and nickel which is 4,000° F (2,204° C).

Inner Core: The innermost core of the earth consists of the heaviest elements of iron and nickel which came to rest here during the early stages of earth's development. These metals are solid despite their temperature of 9,000° F (4,982° C).

The earth is 4,000 miles (6,400 km) in radius; it is impossible to drill into the center since the distance and temperature are too great. Scientists know the structure of the interior of earth through earthquakes. The energy waves given off by earthquakes are like sound waves, traveling faster through solid material than through liquid or gas. There are some earthquake waves which can not pass through a liquid. Knowing this, geologists (people who study the earth) use the speed and direction of earthquake waves which are picked up by instruments around the world to make a picture of the inside of the earth.

The lessons in this section will show how rocks are recycled and what scientists predict about the earth's interior.

30

Recycling Rocks

Question

What is the rock cycle?

Setting the Stage

- Make a transparency of The Rock Cycle (pages 32-33) and use it to explain the rock cycle to students.
- Explain to students that rocks are either igneous, metamorphic, or sedimentary and are constantly being recycled.

Materials Needed for Each Group

- large piece of butcher paper or king-sized white bed sheet
- colored markers or crayons
- copies of Recycling Rocks (pages 32-33), one each per student

   Note to the teacher: The rock cycle will be simulated in a walk through the earth's crust.

Procedure

1. Draw onto the butcher paper or bed sheet an enlarged picture of Recycling Rocks (page 33). Color the igneous rock black, the mantle and lava red, sedimentary rock brown, metamorphic rock various colors, and the ocean water blue. (You may wish to have students help here.)

2. Lay the enlarged drawing on the floor, permitting space for students to walk on it.

3. Demonstrate the walk through the rock cycle while reading aloud the following:
   - This journey begins in the mantle beneath the earth's crust.
   - Magma is forced up, changing to lava and spreading between cracks in a volcano, eventually cooling to become igneous rock. Some of the lava pours out the top and down the sides of the volcano.
   - The lava rolls down the sides of the volcano to the ocean, cooling along the way or when it flows into the water and changes into igneous rock.
   - Wave action breaks the lava and igneous rock into sand-sized pieces.
   - More layers of the sand are added and pressed together, becoming sedimentary rock.
   - More layers of sedimentary rock are formed on top of that layer. Some sedimentary rocks on the bottom get hot because of the pressure and change to metamorphic rock.
   - When the metamorphic rock is buried even deeper, it gets hotter and melts, becoming magma. It may eventually be pushed up again into the crust.

4. Explain to students that this is just one way the rocks in the earth's crust are recycled, for rocks are constantly changing. The rock cycle happens very slowly over millions of years.

5. Have students from each group walk through the rock cycle as you read about their journey.

Extension

Have students collect rocks from the three categories—igneous, metamorphic, and sedimentary—and make a rock cycle on cardboard, showing the direction the cycle moves.

Closure

Have students color in their copy of Recycling Rocks and add it to their geology journals.

Recycling Rocks *(cont.)*

The Rock Cycle

Examples of the Three Types of Rocks

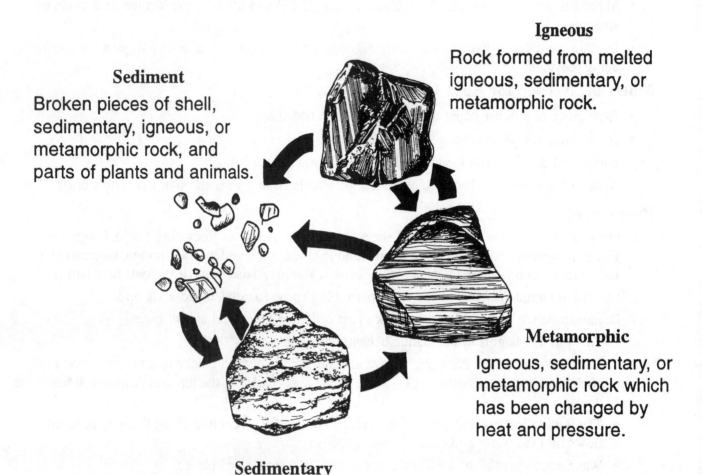

Sediment

Broken pieces of shell, sedimentary, igneous, or metamorphic rock, and parts of plants and animals.

Igneous

Rock formed from melted igneous, sedimentary, or metamorphic rock.

Metamorphic

Igneous, sedimentary, or metamorphic rock which has been changed by heat and pressure.

Sedimentary

Rock made from compressed pieces of sediment.

Examples of the Three Types of Rocks

Igneous—obsidian, pumice, and granite

Metamorphic—marble (from limestone) and slate (from shale)

Sedimentary—sandstone, limestone, and shale

Note: Quartz is found in all three of these types of rocks.

Recycling Rocks *(cont.)*

Recycling the Earth's Crust

Metamorphic rock

Igneous rock

Sedimentary rock

Lava (melted rock)

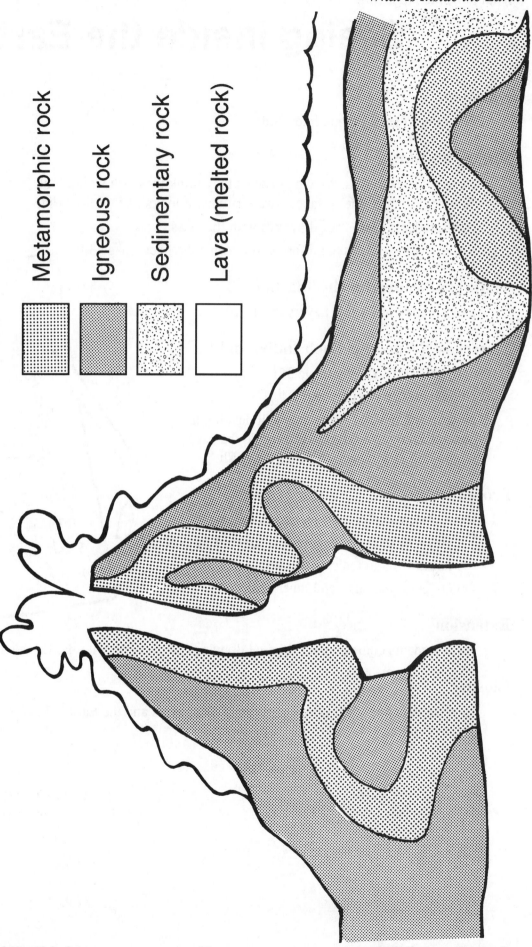

Seeing Inside the Earth

Question

What does the earth look like inside?

Setting the Stage

- Have students make a drawing of what they think the earth would look like if it were cut in half. Tell them to draw the surface as well as what it looks like on the inside.
- Have students share their drawings with the class.
- Review with students the information in Just the Facts (page 30).

Materials Needed for the Teacher

transparency of Earth's Layers (page 35)

Materials Needed for Each Individual

- copy of Earth's Layers (page 35)
- colored markers or crayons

    Note to the teacher: Students will develop an understanding of what it is like inside the earth by looking at a cutaway view of the earth.

Procedure

1. Have students color their picture as follows:
    - Crust—blue, since most of it is covered by oceans
    - Mantle—red to indicate that it is lava
2. Do not color the outer and inner core.

Extension

Have students compare their original drawings of the earth with the Earth's Layers diagram.

Closure

Have students put their drawing of the earth's layers and the Earth's Layers diagram in their geology journals.

Seeing Inside the Earth *(cont.)*

Earth's Layers

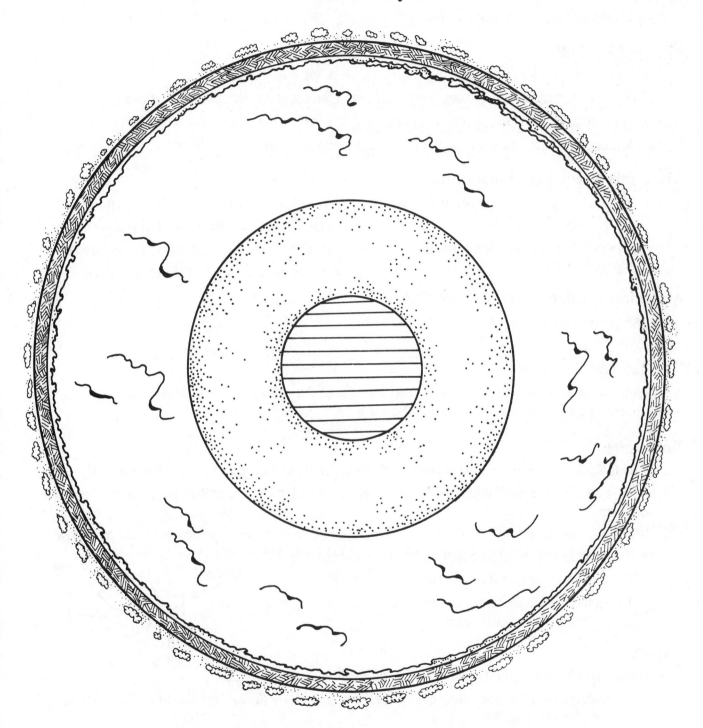

The earth is approximately 8,000 miles (12,800 km) in diameter. This view shows it cut in half so the interior can be seen. Notice that the crust is an extremely thin layer on the surface of the earth. We live on the crust or surface of the earth; nothing lives beneath the crust.

A Scale Model of the Earth

Question

What does the earth look like inside?

Setting the Stage

- Have students look at their drawings of the cutaway view of earth.
- Review with students the various layers and temperatures in the interior of the earth.
- Discuss with students the differences between a flat map and a globe.
- Ask students which one gives us the most accurate view of the world. Why?

Materials Needed for Each Group

- copy of Making a Scale Model of the Earth (page 37), one per student
- three containers of different colored modeling compound (One should be red.)
- piece of waxed paper 12" (30 cm) square
- piece of heavy cardboard 6" (15 cm) square
- pencil, compass, and ruler (metric or standard)
- white paper 8.5" x 5.5" (21 cm x 14 cm)

Materials Needed for the Entire Class

- 18" (45 cm) of plastic fishing line
- two pencils
- dark blue spray paint to paint finished model

Note to the teacher: Students will make a model of the earth's layers to develop an understanding of what it is like inside the earth.

Procedure

1. Divide students into eight groups and distribute the materials they will need for this activity.
2. After students have assembled their earth models, take them outside and spray the dark blue paint on them.
3. When the models are dry, have students use the plastic fishing line to cut them in half as follows:
   - Tie each end of the plastic fishing line to a separate pencil.
   - Encircle the model at its center with the plastic line.
   - Using the pencils as handles, gradually pull the line tighter around the ball until it is cut in half.

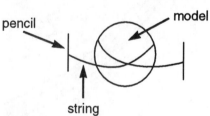

Extensions

- Have students compare their drawings of the earth with their models.
- Have students read the story *How to Dig a Hole to the Other Side of the World* by Faith McNulty. (See bibliography, page 95.)

Closure

Explain to students that the crust of the earth (blue paint) is an extremely thin layer when compared with the rest of the earth. The earth's radius is about 4,000 miles (6,400 km), but the crust is only 5–20 miles (8–32 km) thick.

A Scale Model of the Earth *(cont.)*

Making a Scale Model of the Earth

You will need the following materials for this activity:

- pencil compass and ruler (metric or standard)
- white paper 8.5" x 5.5" (21 cm x 14 cm)
- a piece of waxed paper 12" (30 cm) square
- piece of heavy cardboard 6" (15 cm) square
- three colors (one red) of modeling compound

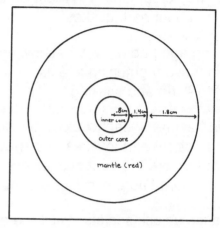

Make a scale drawing of the earth's layers:

- Begin in the center of the paper and use a ruler to draw a line 1.5" (4 cm) long.
- Mark this line off as shown below:

	1.4 cm	
.8 cm		1.8 cm

Scale: .4" (1 cm)= 1,000 miles (1,600 km)
Inner Core: 800 miles (1,280 km) = .3" (.8 cm)
Outer Core: 1,400 miles (2,240 km) = .6" (1.4 cm)
Mantle: 1,800 miles (2,880 km) = .7" (1.8 cm)

- With the cardboard under the paper, place the compass point at the beginning of the line (center of the earth).
- Stretch the compass pencil to the .3" (.8 cm) point on the line. Draw a circle around it.
- Now stretch the pencil to the end of the .6" (1.4 cm) line and draw a second circle.
- Make the last circle by stretching the compass to the end of the .7" (1.8 cm) line.
- Label the parts of your drawing Inner Core, Outer Core, and Mantle.

Follow these simple instructions to make your model:

- Place the piece of waxed paper over the scale drawing you made of the earth's interior.
- Take a piece of the compound you will use for the inner core (save the red for the last layer) and roll it around between your hands until it is a sphere.
- Lay it over the inner core circle to be sure it is the same size. Add or take away extra compound until it fits exactly inside the inner core circle.
- Warm a piece of the second color by squeezing it in your hands. Flatten it out on the waxed paper. Place the inner core in its center, and carefully wrap it around to form a larger sphere. Make it fit inside the outer core circle by adding or taking away compound.
- Use the red compound for the mantle, doing it in layers. Squeeze about 1/4 of the compound until soft. Flatten it into a piece to wrap around the outer core. Continue this until the ball fits exactly inside the mantle circle.

Finishing your model:

The crust of the earth will be represented as blue paint. Your teacher will spray your model with the paint and tell you what to do when it dries.

Just the Facts

Have you ever felt an earthquake? Do you wonder why there are earthquakes? Do you think the earth may break apart? Where do these earthquakes happen? How can we collect information about earthquakes?

People long ago had no way of knowing why earthquakes happened. They made up stories to explain these frightening events. These stories told of gods inside the earth pushing up or holding the earth and shaking it.

California has many earthquakes, but so do other parts of the world. Scientists collect information about earthquakes and show their locations on maps. This helps them learn more about where the earthquakes occur, establishing patterns which can be used for predicting where future earthquakes may happen.

Earthquakes unlock secrets inside the earth. Seismic waves are sent out from the focus of the earthquake in all directions, like a sound wave. These are detected around the world on instruments called *seismographs* which make records of the waves as they arrive. This record is called a *seismogram.* Long waves (L) travel along the crust of the earth and are the slowest, primary waves (P) can pass through the center and are the fastest, and secondary waves (S) go through the earth's interior but reflect off the outer core. The speed of these waves varies with the type and temperature of matter through which they are passing. They travel fastest through a solid, slower through liquid. This information helped seismologists determine which parts of the interior of the earth are solid, liquid, or semisolid, and approximately what the temperatures would be for these sections. The inner core is solid, since P waves travel faster through that area. Scientists have learned that S waves are not able to penetrate liquid matter. When this was discovered, they realized that the outer core must be liquid, since the S waves are diverted when they reach it.

Check the glossary section of this book (pages 93-94) for additional definitions of the terms being used.

The activities in this section will give examples of some of the earthquake legends from around the world. Examples of earthquakes which have happened in California and around the world will be plotted on maps. A working model of a seismograph will be constructed and a simulation will be conducted to illustrate how seismologists locate the epicenter of an earthquake.

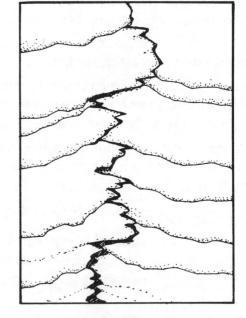

Earthquake Legends

Question

How did people long ago explain the cause of earthquakes?

Setting the Stage

- Tell students to think about an earthquake experience they may have had and answer the following questions:
  - ✓ Where did it happen?
  - ✓ What time of day did it occur?
  - ✓ Where were they when it happened?
  - ✓ What did it feel like?
  - ✓ How did you feel during and after the earthquake?
- After letting students think of answers, have them share with a small group and then select some students to tell the class about their experiences.
- Explain to students that although geology (the study of the earth) was developing in the 19th century, it was not until the late 1960's that scientists were able to explain the cause of earthquakes. Long ago, people who lived in areas where earthquakes took place often made up stories to explain them.
- Tell students that this lesson will introduce them to some earthquake legends and let them write one of their own.

Materials Needed for Each Individual

- white construction paper
- crayons or colored markers
- copy of Examples of Earthquake Legends (page 40)

  Note to the teacher: Before scientists learned of the cause for earthquakes, people made up stories to explain this phenomenon. Several examples of these stories should be shared. Then students should write a brief description for why earthquakes happen and draw a picture to illustrate their story.

Procedure

1. Read to students Examples of Earthquake Legends. Encourage them to close their eyes and imagine a picture which would go with the stories as you read them.
2. Have students locate the sites of the stories on a world map.
3. Ask students to make a picture which will show why earthquakes occur.
4. Have students write a brief story on the back of their pictures to explain earthquakes.

Extensions

- Divide students into small groups and let them share their stories and pictures.
- Have students display their pictures on a classroom bulletin board.

Closure

After student pictures have been displayed in the classroom, have them include their pictures and stories in their geology journals.

Earthquake Legends *(cont.)*

Examples of Earthquake Legends

Assam (between Bangladesh and China)

A race of people live inside the earth, and sometimes they shake the ground to find out if there are any people living on the surface. When children feel a quake, they shout "Alive! Alive!" to let the people inside know they can stop shaking the ground.

Columbia

When the earth was first made, it rested firmly on three large beams of wood. One day, the god Chibchacum decided that it would be fun to see the plain of Bogotá underwater. He flooded the land, and for his punishment he is forced to carry the world on his shoulders. When he gets angry, he stomps his foot, shaking the earth.

Greece

According to the Greek philosopher Aristotle, strong winds are trapped and held in underground caverns. Earthquakes are caused by their struggle to escape.

India

The Hindus thought that the earth was supported by four elephants, the strong animals which they used to do heavy work. The elephants stood on the back of a turtle which pulled the earth through celestial "waters." When any of these animals moved, the earth would tremble and shake.

Japan

A huge catfish is curled up under the ocean, with the island of Japan resting on his back. A demigod (part god, part human) holds a heavy stone over the catfish to keep it still. Sometimes the demigod gets tired and looks away, the catfish moves, and Japan shakes.

Mexico

When *El Diablo*, the devil, wants to stir up trouble for people on earth, he makes giant rips in the ground from inside the earth.

Tennessee, USA

The Indians tell a legend of a Chickasaw chief who fell in love with a beautiful Choctaw princess. He was young and handsome but had a twisted foot, so his people named him Reelfoot. When the princess' father refused to let Reelfoot marry his daughter, the chief and his friends kidnapped her and held a wedding. The Great Spirit was angry and stomped his foot. The shock made the Mississippi River overflow its banks and drown the wedding party. (Reelfoot Lake on the Tennessee side of the Mississippi actually formed as a result of the New Madrid earthquake of 1812.)

Siberia

The earth rides on a sled driven by a god named Tuli and pulled through the heavens by dogs. The dogs have fleas which make them stop to scratch, causing the earth to shake.

California Earthquakes

Question

Where do earthquakes happen in California?

Setting the Stage

- Ask students if they know what part of the United States has many of its earthquakes. (Most will respond California.)
- Tell students that in this lesson they will plot some of the California earthquakes on a map of the state to see where they occur.

Materials Needed for Each Group

- road map of California which includes an index of towns and cities
- 54 red self-adhesive dots
- two thin cardboard strips or yard (meter) sticks to use as guides on the map (optional)
- data-capture sheet (page 42), one per student

   Note to the teacher: Two maps which will enhance this lesson are available through the California Division of Mines and Geology, Publication Sales, P.O. Box 2880, Sacramento, CA 95812. Orders must be accompanied by a check. These maps are as follows:

   *#1 Fault Map of California by C.W. Jennings, 1975, about $7.00 (Request a folded copy.)

   *MS 39 Earthquake Epicenter Map of California, by D.L. Parke, et. al. 1976, about $3.00

   Students will plot the location of earthquakes on a California map and analyze this data to see if they can find any patterns.

Procedure

1. Divide students into eight groups and provide each of them with the materials to do this activity.
2. Let students examine the California map to familiarize themselves with the coordinates.
3. Have students use the index on the map to find and record the coordinates for each location.
4. Show students how to plot the first piece of data so they will be able to use the coordinate system.
5. Be sure students take turns plotting each earthquake site.

Extension

Project for students a transparency copy of Major Earthquake Faults for California and discuss the locations of the major faults shown on this map. Explain that this map is the result of plotting many years of data which showed where earthquakes occurred. When these areas are linked together, major faults are outlined.

Closure

Have students look at the maps when they finish and discuss which areas of California have the most earthquakes. Point out that this data is very limited and that thousands of earthquake locations need to be plotted before geologists can begin to draw maps which show the areas where these earthquakes are most likely to occur.

California Earthquakes *(cont.)*

California Earthquake Epicenters
Magnitude 5.5+
1812–1983

Note to the teacher:

The following data have been sorted by location on the California map from north to south, to make it easier for students to plot the information on the map. (Cover up before giving to students.) The coordinates for each area can be found in the index of the map and then placed in the column labeled Coor. Place the dots near each other in areas where more than one earthquake occurred.

Coor.	Epicenter	Year	Mag.
	Crescent City	1962	5.6
	Eureka	1932	6.4
	Eureka	1954	6.6
	West of Ferndale	1956	6.0
	Ferndale	1967	5.8
	Fortuna	1975	5.7
	West of Arcata	1960	5.7
	Lassen Park	1950	5.5
	Chico	1940	6.0
	Loyalton	1959	5.6
	Off Cape Mendocino	1951	6.0
	Off Cape Mendocino	1954	6.2
	Oroville	1975	5.8
	Santa Rosa	1969	5.7
	San Francisco	1906	8.3
	San Francisco	1938	7.9
	Hayward	1968	7.5
	Livermore	1980	5.5
	Mammoth Lake	1980	6.1
	Bishop	1927	6.0
	Bishop	1978	5.8
	Coyote	1911	6.6

Coor.	Epicenter	Year	Mag.
	East of San Jose	1955	5.8
	Lone Pine	1872	8.1
	Monterey Bay	1926	6.1
	Gilroy	1979	5.9
	Hollister	1961	5.6
	Coalinga	1983	6.5
	Parkfield	1934	6.0
	Shandon	1922	6.5
	Bakersfield	1952	7.7
	Bakersfield	1952	5.8
	Maricopa	1954	5.9
	East of Barstow	1947	6.4
	Lompoc	1812	7.9
	San Fernando	1971	6.4
	San Bernardino	1923	6.2
	Twenty-nine Palms	1949	5.9
	Santa Barbara	1925	6.2
	Santa Barbara	1941	5.9
	Santa Barbara	1978	5.7
	Oxnard	1973	5.9
	Long Beach	1933	6.3
	San Jacinto	1918	6.8

Source: **California Earthquake Education Project (CALEEP), 1986**

Earthquakes Around the World

Question

Where do earthquakes occur in the world?

Setting the Stage

- Review with students California Earthquakes (pages 41-42) and show one of the maps on which earthquake epicenters were plotted.
- Ask students if they know of any other parts of the world which have earthquakes.
- Tell students that they will plot sample information gathered for areas around the world which had earthquakes of a magnitude of 6.1–6.8 between January 1983 and December 1985.

Teacher Preparation

- Find a large world map which clearly shows latitude and longitude. Use a world map from your school or purchase one from the American Automobile Association (AAA).
- The World Earthquake Epicenters data has been sorted by longitude to make plotting this information easier for the students. There are 98 sites in all which should be divided among the students. Place the data on file cards, using three to four sites per card in the order they are listed on the data sheet.

Materials Needed for Each Individual

- file cards showing three epicenters
- self-adhesive red dots

  Note to the teacher: Students will plot earthquake epicenters from around the world on a world map.

  The Dynamic Planet, Volcanoes, Earthquakes, and *Plate Tectonics* is an excellent map to use with this activity. The map is 60" x 40" (150 cm x 100 cm). It shows the earth's plates and the location of many earthquake epicenters and volcanic eruptions. It is available from the National Science Teachers Assoc., 1840 Wilson Blvd., Arlington, VA 22202-3000.

Procedure

1. Familiarize students with the world map and the latitude and longitude grid.
2. Place a red dot on the map to show the first piece of data so students can see how to use the coordinate grid.
3. Let each student plot three sites.
4. Have students place the dots near each other in areas where more than one earthquake occurred.
5. It will take time to carefully plot each site. Some students may plot the data while others are doing another activity such as viewing a film about earthquakes.

Extensions

- Have students research why earthquakes occur most commonly where they do.
- Have students present their earthquake research to the class.

Closure

In their geology journals, have students list where the earthquakes occur. They will notice that most occur around the rim of the Pacific Ocean, with some clustering near India.

Earthquakes Around the World *(cont.)*

World Earthquake Epicenters
1983–1985
Magnitudes 6.1–6.8

Note to the teacher: This data has been sorted by longitude to make it easier to plot on the map. Cut out coordinates in groups of three and give one group to each individual.

° Latitude	° Longitude		° Latitude	° Longitude		° Latitude	° Longitude
22 N	103 E		54 N	158 E		61 N	147 W
9 S	114 E		10 S	160 E		58 S	16 W
18 N	121 E		56 N	161 E		55 N	160 W

° Latitude	° Longitude		° Latitude	° Longitude		° Latitude	° Longitude
24 N	122 E		54 N	161 E		15 S	173 W
20 N	122 E		56 N	163 E		51 N	174 W
0	123 E		5 S	167 E		16 S	174 W

° Latitude	° Longitude		° Latitude	° Longitude		° Latitude	° Longitude
8 N	124 E		23 S	171 E		15 S	174 W
5 N	125 E		22 S	171 E		21 S	175 W
2 N	126 E		18 S	172 E		17 S	175 W

° Latitude	° Longitude		° Latitude	° Longitude		° Latitude	° Longitude
7 S	128 E		37 S	177 E		26 S	176 W
7 N	130 E		38 N	20 E		19 S	178 W
28 N	130 E		40 N	25 E		21 S	179 W

° Latitude	° Longitude		° Latitude	° Longitude		° Latitude	° Longitude
34 N	131 E		40 N	42 E		35 S	179 W
32 N	132 E		41 N	62 E		35 S	179 W
8 N	137 E		40 N	63 E		60 S	25 W

° Latitude	° Longitude		° Latitude	° Longitude		° Latitude	° Longitude
34 N	137 E		19 N	67 E		1 N	28 W
31 S	138 E		37 N	71 E		8 N	39 W
40 N	139 E		36 N	71 E		61 S	53 W

Source: U.S. Geological Survey, Golden Colorado

Earthquakes Around the World *(cont.)*

World Earthquake Epicenters
1983–1985
Magnitudes 6.1–6.8

Note to the teacher: This data has been sorted by longitude to make it easier to plot on the map. Cut out coordinates in groups of three and give one group to each individual.

° Latitude	° Longitude
40 N	139 E
41 N	139 E
29 N	139 E

° Latitude	° Longitude
36 N	71 E
7 S	72 E
39 N	75 E

° Latitude	° Longitude
17 N	62 W
28 S	63 W
27 S	71 W

° Latitude	° Longitude
40 N	140 E
34 N	142 E
46 N	144 E

° Latitude	° Longitude
50 N	79 E
13 N	94 E
6 N	95 E

° Latitude	° Longitude
31 S	71 W
33 S	72 W
40 S	76 W

° Latitude	° Longitude
44 N	147 E
44 N	148 E
49 N	151 E

° Latitude	° Longitude
0	98 E
18 N	103 W
44 N	114 W

° Latitude	° Longitude
28 N	78 W
9 S	79 W
0	80 W

° Latitude	° Longitude
45 N	151 E
10 S	151 E
36 N	153 E

° Latitude	° Longitude
5 S	120 W
15 N	121 E
62 N	124 W

° Latitude	° Longitude
10 N	80 W
9 N	83 W
15 N	94 W

° Latitude	° Longitude
4 S	153 E
5 S	154 E
15 S	156 E

° Latitude	° Longitude
41 N	127 W
44 N	128 W
12 N	14 W

° Latitude	° Longitude
16 N	95 W
19 N	155 W

Source: U.S. Geological Survey, Golden Colorado

Earthquake Waves

Question

> How is a seismogram made?

Setting the Stage

- Describe to students how a record of an earthquake is made by a seismograph by using the information in Just the Facts (page 38) and transparencies or copies of the following:

  > Seismogram: Aftershock of Northridge, CA Earthquake (page 47)

  > Seismic Waves (page 48)

  > Seismographs (page 49)

- Tell students that they will construct a model of a seismograph and make a seismogram.

Materials Needed for Each Group

- copy of Building a Simple Seismograph (page 50), one per student
- oatmeal box or 1 lb (500 g) coffee can with holes drilled in each end
- shoe box with a hole in each end and large enough to hold the oatmeal box or can
- sharpened pencil
- .5" (1.25 cm) wooden dowel about 18" (46 cm) long
- white paper large enough to cover oatmeal box or coffee can
- masking and transparent tape

  Note to the teacher: Students will make a working model of a seismograph and create a seismogram.

Teacher Preparation

- Drill holes in both ends of the shoe box and oatmeal box or coffee can, large enough for the dowel to pass through both the shoe box and oatmeal box or coffee can (drum.)
- Cut away one long side of the shoe box to permit access to the rotation drum. Leave about 1" (2.5 cm) of border around this cut for additional support.

Procedure

1. Have students follow the instructions for Building a Seismograph (page 50).
2. Monitor students and offer assistance if needed. Be sure they rotate responsibilities.

Extensions

- Have students contact their local science museum or geology department of a college or university to receive a copy of recent seismograms.
- Have students share their seismograms with the rest of the class and explain where the Primary (P), Secondary (S), and Long (L) waves are shown.

Closure

> Have students place copies of seismograms and their explanations in their geology journals.

Earthquake Waves *(cont.)*

Seismogram: Aftershock from Northridge, CA Earthquake

On January 17, 1994, a 6.6 magnitude earthquake occurred about 30 miles (48 km) below the area of Northridge, CA. This seismogram was made 120 miles (192 km) south on a seismograph at the San Diego Natural History Museum. It shows an aftershock which occurred at approximately 3:30 p.m. on January 17, 1994. There were thousands of aftershocks from this earthquake, some nearly as strong as the first one. The first wave detected was the primary (P) wave, followed by the secondary (S) wave about 10 seconds later, and then the long (L) wave.

As the drum of the seismograph rotates, the pen draws a continuous line around it. Every minute the pen receives a pulse which makes it jerk, leaving a "blip." These have been marked on this seismogram with numbers 1-3.

The drum takes 15 minutes to rotate; thus each line is 15 minutes after the line above. When an earthquake occurs, the pen receives an electrical impulse from a sensor which is embedded in bedrock below the museum. The pen moves in a side-to-side motion, with the strokes varying in size depending upon the strength (magnitude) of the earthquake.

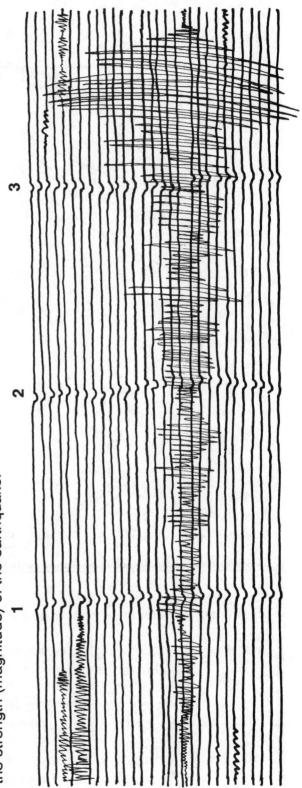

Earthquake Waves *(cont.)*

Seismic Waves

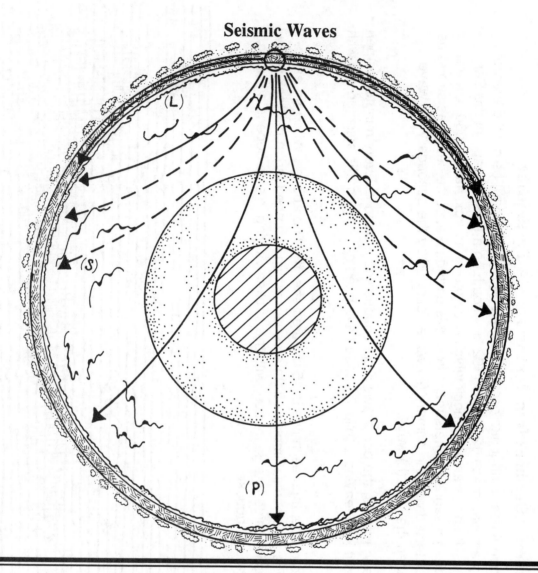

	Key
Epicenter:	center of earthquake on the crust.
Focus:	location of actual earthquake which may be miles (km) below the epicenter but still in the earth's crust.
(P)	Primary waves travel the fastest, pass through the earth, and are the first to be detected by seismographs.
(S)	Secondary waves move slower, will not pass through the liquid outer core, and are the second seismic waves to reach the seismographs.
(L)	Long waves travel out from the focus through the earth's crust and take the longest time to reach the seismograph.

Earthquake Waves (cont.)

Seismographs

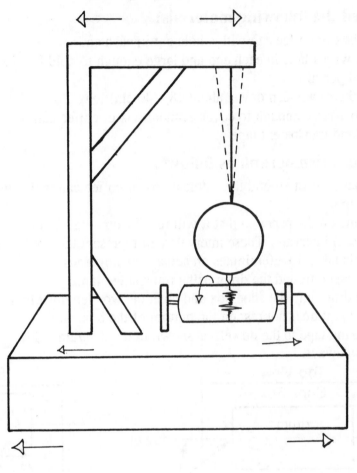

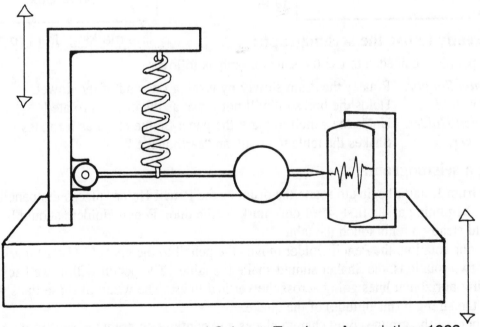

Source: *Earthquakes,* National Science Teachers Association, 1988

Earthquake Waves *(cont.)*

Building a Simple Seismograph

You will need the following materials:

- oatmeal box or coffee can with a hole cut in each end
- shoe box with a hole in each end and large enough to hold box or can
- sharpened pencil
- 1/2" (1.25 cm) wooden dowel about 18" (46 cm) long
- white paper large enough to cover oatmeal box or coffee can
- masking and transparent tape

Construct the seismograph as follows:

- Cut the paper to fit around the oatmeal box or coffee can which will be the drum of your seismograph.
- Draw a line on the paper so that it will reach from one end of the drum to the other. Mark it off in .4" (1 cm) intervals. These intervals will represent the time it takes for the drum to rotate once. This takes 15–30 minutes on actual seismographs.
- Tape the paper around the drum using transparent tape.
- Place the drum into the shoe box and push the wooden dowel through the side of the box and drum so that it does not rest on the bottom of the box.
- Add masking tape to the dowels as shown in the diagram below. This will prevent the drum from moving sideways.

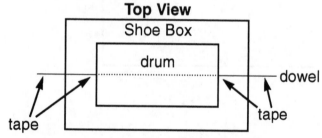

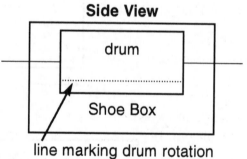

Getting ready to use the seismograph:

Four people are needed to use the seismograph as follows:

Drum Rotator: Rotates the drum slowly by turning one end of the dowel.
Box Holder: Holds the box so it will not move as the drum is rotated.
Pencil Holder: Holds the pencil to mark the paper on the drum as it rotates.
Shaker: Shakes the table to create an "earthquake."

Drawing a seismogram:

- The Drum Rotator begins to rotate the drum as the Pencil Holder places the pencil against the drum, beginning at the first .4" (1 cm) mark on the line. Pencil Holders should brace their arms to hold steady but not touch the table.
- After one rotation, the Pencil Holder moves the pencil to the next .4" (1 cm) mark.
- After two rotations, the Shaker should shake the table. The pencil will move back and forth, drawing horizontal lines going across the vertical lines. The width of these lines will depend upon the strength (magnitude) of the quake.
- Members of the group should change places and continue to draw the seismogram until the drum is filled, alternating between quiet and earthquake periods.

50

Locating an Earthquake

Question

How do scientists know where earthquakes occur?

Setting the Stage

- Have students discuss what they have learned about seismographs and seismograms.
- Explain to students that this lesson will enable them to see how seismographs are used to determine the exact location of an earthquake.
- Introduce to students the term *triangulation* and discuss its meaning and uses.

Materials Needed for Each Group

- large map of the United States with scale of miles (km) on it.
- one push-pull pin
- three pieces of string, each 12" (30 cm) long
- thick piece of cardboard 4" (10 cm) square
- red pen or self-adhesive red dot
- pencil
- data-capture sheet (page 52), one per student

    Note to the teacher: Students will learn to use the method of triangulation to find the epicenter of an "earthquake."

Procedure

1. Divide students into eight groups and distribute the materials needed.
2. Let students look at their maps and become familiar with them. Help them to locate the following items:
    ✓ scale of miles (km)
    ✓ Chicago, Dallas, and San Francisco
3. Demonstrate to students how to make a circle using string, a pencil, and the pin as shown below:

Explain to students that the pin must be pushed through the end of the string, the map, and into the piece of cardboard to hold it securely. The pencil point is placed into the loop of the string and then pulled tight to stretch the string to its full length. A circle is drawn using the string to guide the point of the pencil.

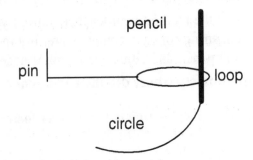

Extension

Have students share the result of this activity. They should have found that the epicenter of the earthquake is where the three circles overlap; in this case it was near New Madrid, Missouri.

Closure

Have students complete their data-capture sheets and add them to their geology journals.

Locating an Earthquake *(cont.)*

Where Was That Earthquake?

You are a seismologist (a person who studies earthquakes) working at a center in Chicago which gathers earthquake information. The seismograph has just recorded an earthquake. The arrival time between the primary and secondary waves indicates the distance of the earthquake from Chicago was 375 miles (600 km). You need to show where the epicenter of the earthquake was.

Materials Needed

- large map of the United States with scale of miles (km) on it.
- one push-pull pin
- three pieces of string, each 12" (30 cm) long
- thick piece of cardboard 4" (10 cm) square
- red pen or self-adhesive red dot
- pencil

Procedure *(Student Instructions)*

1. The earthquake could have taken place anywhere within a 375 mile (600 km) radius of Chicago. You will need two other seismic readings to locate the exact epicenter of the earthquake. The seismic readings shown below were received from other seismographs.

Location of Seismograph	Miles from Earthquake
Dallas	560 miles (896 km)
San Francisco	900 miles (1,440 km)

2. You will need to show the distance from each of the three centers to the earthquake. This must be done with a circle around each center, having a radius which is equal to the distance in miles (km) from the receiving center to the earthquake. A circle is being used rather than a straight line, since the information tells only the distance, not the direction of the earthquake.

3. Locate the scale of miles (km) on the map you are using.

4. Use the method your teacher showed you for making a circle with string. Be sure to use a different string for each center. Carefully measure the string when it is stretched tight between the pin and pencil and adjust the length if necessary to show the correct miles (km).

Approximately where did the earthquake happen? _____

How do you know this is the correct location? _____

_____

_____

Just the Facts

Why are there earthquakes? What could crack the earth's crust? What makes mountains and volcanoes? How do the continents drift?

The history of the earth is literally written in stone: fossils and rock formations. These can be interpreted by scientists to confirm what once was thought impossible—that huge sections of the earth's crust drift on a sea of molten magma, carrying continents and the ocean floor with them. Large land animal fossils and rock layers found along the eastern coast of South America and western coast of Africa were found to be the same. Fossil specimens found in Australia and Antarctica also indicate that the same type of land animals once lived in these widely separated areas. Plant fossils show that climates in many parts of the world have gradually changed.

Drilling samples of rock, mapping contours of the ocean floor, and plotting earthquake and volcano locations have revealed outlines of large divisions in the earth's crust. This evidence, along with observations of volcanic activity deep under the ocean, has proven the early theories of continental drift and sea floor spreading. The theory is now referred to as *Plate Tectonics* and explains how continents gradually relocate over the earth's long history. It also predicts what will happen to the crust as the energy from within the earth's mantle continues to push the plates around, creating earthquakes, building and destroying mountains and volcanoes, and gradually changing the crust.

The activities in this section will illustrate movement of the crust and the mechanism which scientists think makes it move. Using peanut butter and jelly sandwiches, hard boiled eggs, and boiling water, students will develop an understanding of plate tectonics.

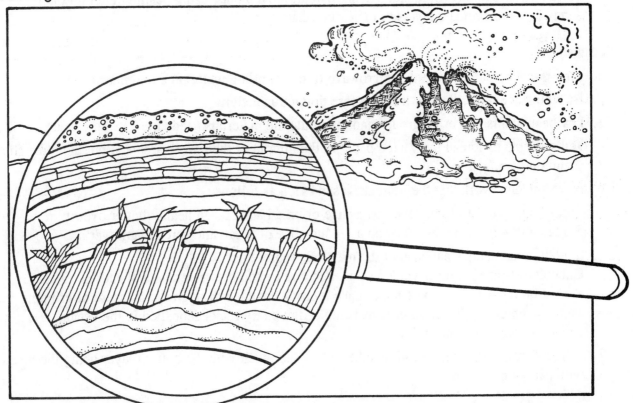

Shifting Crust

Question

Does the crust move?

Setting the Stage

- Remind students of the walk through the earth's crust they did in an earlier activity (pages 31-33).
- Review with students the make up of the earth's crust by showing the transparency of Recycling the Earth's Crust (page 33) or the large drawing. Mention the three types of rocks which are constantly being recycled.
- Explain to students that this activity will show them more about the earth's crust and how it moves.

Materials Needed for Each Group

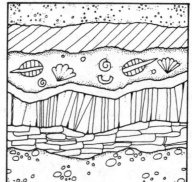

- three pieces of fresh whole wheat bread
- strawberry jam
- smooth peanut butter
- 2 tbs (30 g) of melted plain chocolate bar
- plastic spreading knife
- waxed paper and napkin or paper towel
- copy of The Movement of Earth's Crust (page 56), one per student
- data-capture sheet (page 57), one per student

   Note to the teacher: Students will show fault movements, using peanut butter and jelly sandwiches to simulate layers of the earth's crust.

Procedure

1. Melt the chocolate bars in the microwave or over boiling water on a hot plate.
2. Be sure students wash their hands before doing this activity.
3. Divide students into eight groups and distribute the materials needed.
4. Tell students to spread the waxed paper in the middle of their work area and put the three slices of bread on the paper.
5. Have students work together to make a sandwich as follows:

   Spread a thin layer of chocolate on a slice of bread and spread peanut butter over it.
   Spread a thin layer of chocolate on a second slice of bread and spread jam over it.
6. Explain to students what each of the materials represents:
   Chocolate—igneous rock which was melted and forced between layers of rock
   Peanut butter—metamorphic rock, made by rock changed through the pressure of grinding
   Jam—sedimentary layer, with seeds being sedimentary rocks deposited in the ocean
   Bread—more sedimentary layers
7. Tell students to put their bread together into a sandwich with the bottom layer being the bread with peanut butter on it.

Shifting Crust *(cont.)*

8. Have students draw on their data-capture sheets the three types of faults using their peanut butter and jelly sandwiches as a model.

Extensions

- Have students slice each of the sandwiches so everyone will get two sections showing the layers.
- Using a transparency of The Movement of Earth's Crust (page 56), explain to students the three main types of faults which cause earthquakes.
- Have students move their sandwich slices to represent the faults and to see the results.
- Take students to a road or stream bed to see the layers which are exposed. Have them look for evidence of layers which have been pushed up above layers which are next to them.

Closure

Have students complete their data-capture sheets and add them to their geology journals. Also have students write about what might happen to roads and rivers which run on top of the crust as faults move.

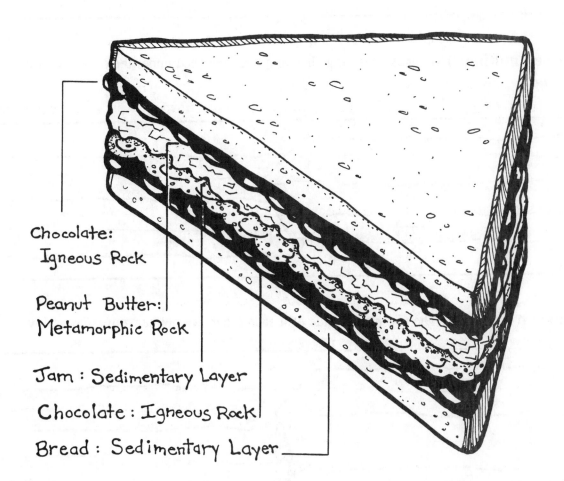

Chocolate: Igneous Rock

Peanut Butter: Metamorphic Rock

Jam: Sedimentary Layer

Chocolate: Igneous Rock

Bread: Sedimentary Layer

Shifting Crust *(cont.)*

The Movement of the Earth's Crust

The movement of the earth's crust is called *faulting*. This movement of the crust causes earthquakes. The three main types of faults are normal faulting, reverse faulting, and strike-slip faulting. All three are shown below.

Normal Faulting: Two blocks of the earth's crust move apart from each other. One block may also drop below the other.

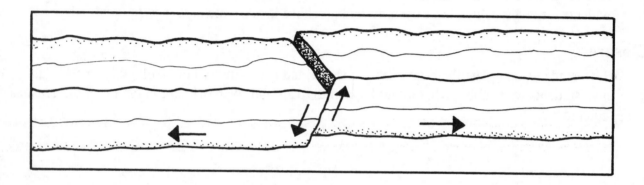

Reverse Faulting: Two blocks push together, and one is pushed under the other. This is called *subduction*.

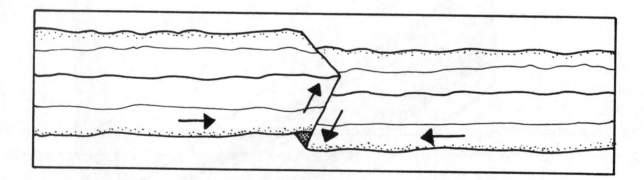

Strike-slip Faulting: The two blocks slide past each other, but neither moves up or down.

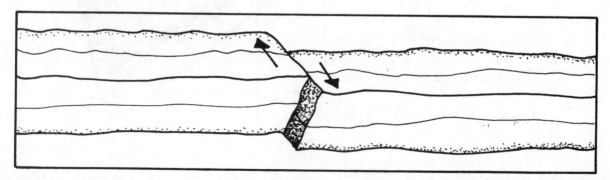

Shifting Crust *(cont.)*

Draw and label each of the three types of faults using your peanut butter and jelly sandwich as a model.

Broken Egg

Question

How is the earth like a hard boiled egg?

Setting the Stage

- Review with students the layers of the earth by projecting a transparency of Earth's Layers (page 60).
- Tell students that the earth is much like a hard boiled egg as they will see in this lesson.

Materials Needed for Each Group

- hard boiled egg
- black felt tip marker
- copy of Earth's Plates (page 59), one per student
- copy of Earth's Layers (page 60), one per student

Procedure

1. Divide students into groups and provide them with their materials.
2. Tell students the earth's crust is much like an egg shell, very thin and brittle.
3. Have students crack the shell all around the edge by lightly tapping it on a table. They should not peel off the shell.
4. Ask students to notice how the shell is broken and that some pieces overlap others.
5. Have students use a felt tip pen to carefully outline the edges of the cracked pieces.
6. Explain to students that this is just like the earth's crust by showing a transparency of Earth's Plates (page 59), telling them that these sections of the crust are referred to as *plates*.

Extensions

- Cut the eggs in half and have each group place a black dot about the size of a pea into the center of the yolk. This represents the inner core of the earth. Using the Earth's Layers (page 35), have students look at the thickness of the shell by comparison to that of the rest of the egg and explain that the earth's crust is actually much thinner, relative to the rest of the earth.
- Have students demonstrate to someone at home how the hard boiled egg is like the earth. This will enable them to further develop their own understanding of this concept.

The Earth is Like an Egg.

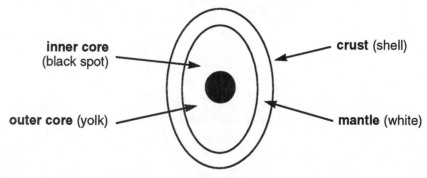

Closure

In their geology journals, have students write a story about what it would be like to travel to the center of the earth.

Broken Egg *(cont.)*

Earth's Plates

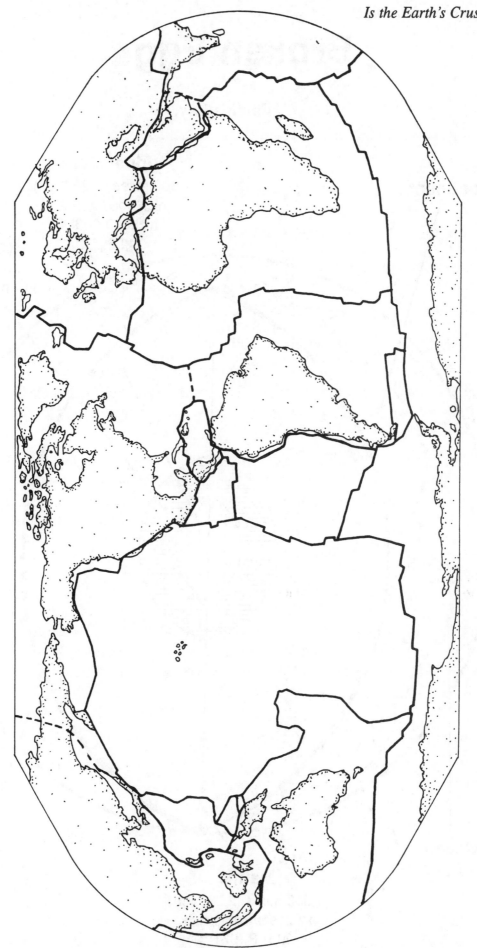

Broken Egg *(cont.)*

Earth's Layers

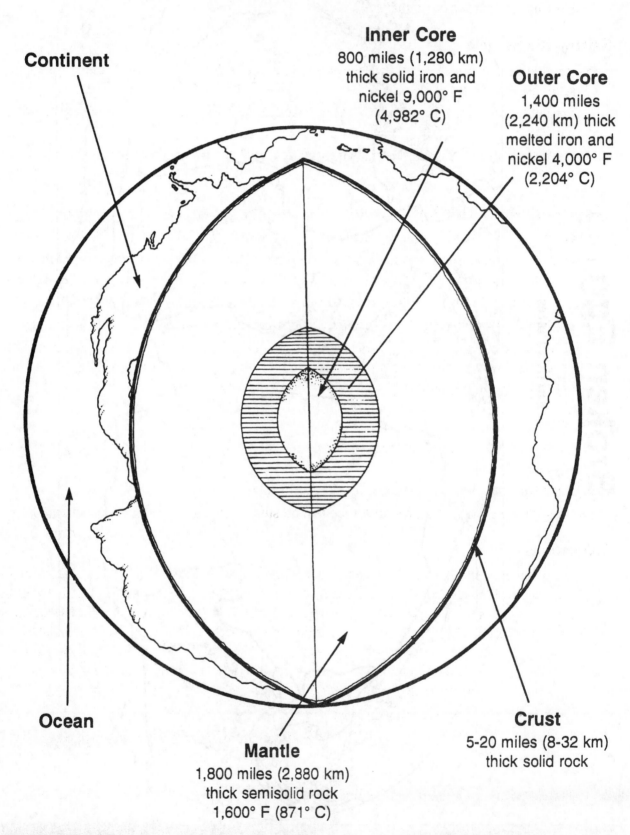

Continent

Inner Core
800 miles (1,280 km)
thick solid iron and
nickel 9,000° F
(4,982° C)

Outer Core
1,400 miles
(2,240 km) thick
melted iron and
nickel 4,000° F
(2,204° C)

Ocean

Crust
5-20 miles (8-32 km)
thick solid rock

Mantle
1,800 miles (2,880 km)
thick semisolid rock
1,600° F (871° C)

Cracked Earth

Question

How has the earth's crust changed?

Setting the Stage

- Remind students of the Broken Egg (page 58) and how the thin crust of earth is broken into sections.
- Project for students a transparency of the Cracked Earth Map (page 62) and review the different parts.
- Have students use this map to place information collected by scientists to prove the location of the plates.

Materials Needed for Each Group

- copy of the Cracked Earth Map (page 62), one per student
- copy of the maps showing the earth's changes from 200 million years ago to 50 million years in the future (pages 23-28), one each per student
- red, green, and brown markers or crayons
- copy of the Continents Adrift Worksheet (page 63), one per student

Note to the teacher: Students will learn the location of the plates which make up the earth's crust and how scientists have proven the theory of plate tectonics. They will also relate these plates to the shifting of the continents over time.

Procedure

1. Have students examine the Cracked Earth Map (page 62) and study the location of the major plates.
2. Have students complete their Continents Adrift Worksheets (page 63).
3. Monitor the groups as they work to see that all students participate and ask questions.

Extensions

- Lead students through a comparison of where the plates are located and the changes which have occurred in the earth's crust during the past 200 million years. Each group should spread their maps in order in front of them and place the Cracked Earth Map near them. Have students explain what was happening to the plates during these years. They should realize that the Atlantic Ocean floor was spreading apart along the mid-Atlantic ridge. The continents were carried on top of the moving plates to their present locations. The plates will continue to move, and thus the continents will be in a different location 50 million years from now.
- Tell students that the drawings of the past and future locations of the continents are based upon the geological evidence that we have today. They may change as we learn more about our earth.

Closure

Have students complete their Continents Adrift Worksheets and add them to their geology journals.

Is the Earth's Crust Cracked?

Cracked Earth *(cont.)*

Cracked Earth Map

Most earthquakes occur near or along the boundaries of the rocky plates of the earth's crust. Each dot on this map shows where a major earthquake has occurred during the past 30 years. The arrows indicate the direction of the plates' movement.

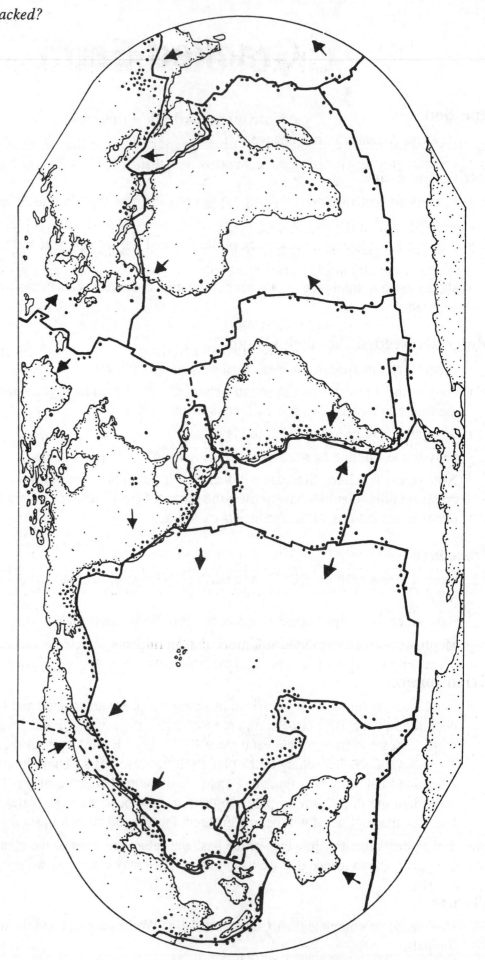

Cracked Earth *(cont.)*

Continents Adrift Worksheet

You are scientists who are examining the evidence to try to prove that the earth's crust is divided into plates and that they move, taking the continents with them.

Materials You Will Need

- enlarged copy of the Cracked Earth Map
- copies of the maps showing the earth's changes from 200 million years ago to 50 million years in the future
- red, green, and brown markers or crayons
- Use the scientific data collected over many years to color code the Cracked Earth Map as directed.

Scientific Evidence	Code
The same fossils of large land animals are found on the eastern edge of South America and the western edge of Africa.	Green line along the east coast of South America and west coast of Africa
The same type and age of rock formations are found on the eastern edge of South America and the western edge of Africa.	Brown line along the east coast of South America and west coast of Africa
Fossil remains of animals like those found in Australia have been discovered in Antarctica.	Green X's on both of these continents
A long, high ridge runs north–south along the middle of the floor of the Atlantic Ocean. Rock samples taken along the top of this ridge show the rocks are younger than those taken further away from the ridge.	Red along the mid-Atlantic ridge. Brown on either side of the ridges
Deep sea diving vessels investigating areas along the ridges discover molten magma pushing up. This magma cools and becomes new rock.	Red along west and south edge of the Nazca Plate
The mountains which run along the western coast of North and South America are gradually rising higher. There are frequent earthquakes in this area also. The mountains on these coasts also contain active volcanoes, like Mount St. Helens, as well as many which are now dormant but not extinct.	Red along the western coasts of North and South America
Earthquakes happen frequently along and near the San Andreas Fault, located in California. This fault outlines the division between the Pacific and North American Plates.	Red along the edge of the Pacific Plate which divides California and Baja from the North American Plate
The Himalaya Mountains in northern India rise higher each year. There are many earthquakes here also.	Red along the division between India and Asia

Moving Machine

Question

What makes the crustal plates move?

Setting the Stage

- Show students the Cracked Earth Map (page 62) and remind them of the activity in which they compared the location of the continents over 200 million years.
- Review with students the scientific discoveries which proved that the earth's crust is divided into plates.
- Review with students the information on the transparency of Earth's Layers (page 60).
- Tell students you are going to do a demonstration which will show how scientists think these plates can move.

Materials Needed for Each Individual

- colored crayons or markers
- copy of Convection Currents Move the Crust (page 66)
- data-capture sheet (page 67)

Materials Needed for Teacher Demonstration

- large glass bread dish (oven safe)
- aquarium heater
- red and blue food coloring
- ice in a small sealable plastic bag
- water
- masking tape

Note to the teacher: A demonstration of convection currents will be done by the teacher to simulate the method scientists think moves the crustal plates of the earth.

Procedure

1. Fill a glass bread dish with water leaving it 1" (2.5 cm) from the top.
2. Set bread dish high enough so that all students can see through the glass sides.
3. With masking tape, anchor the aquarium heater perpendicularly at one end of the container.
4. Let the heater warm the water for at least five minutes before adding the bag of ice.
5. Place the bag of ice into the water at the other end of the container and secure it with tape.
6. As your students watch closely, carefully place a drop of red food coloring near the heater. Tell them this represents hot water. Let them observe where the water goes. (It stays on the surface and spreads out toward the other end of the container.)
7. After the red has spread to the ice, have students watch to see what happens. (It begins to drop toward the bottom when it reaches the ice.)
8. Add a drop of blue food coloring near the ice and observe it. The coloring sinks and begins to spread out toward the heater. When it reaches the heater, it begins to rise.
9. Observe the colored water until all students realize that hot water will always rise until it cools, and then it will sink.

Moving Machine *(cont.)*

Procedure *(cont.)*

10. Have students draw what they observe on their data-capture sheets.

Extensions

*   Have students investigate the properties of hot and cold liquids. Provide them with hot water which has been colored red with food coloring and cold water colored with blue food coloring. Have them use eyedroppers to put drops of the liquids into clear water which is at room temperature. They will see that the hot water will remain on top and the cold water will sink to the bottom.

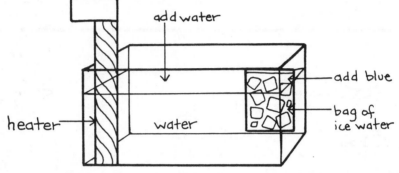

*   Explain that a convection current is set up as the water is heated and it becomes less dense. It thus rises to the surface where it is cooled and becomes more dense, thus sinking. This continues as long as the sources of heat and cold remain.

*   Compare this to what happens in the earth. Use the transparencies of Earth's Layers (page 60) and Convection Currents Move the Crust (page 66) to explain how this motion occurs inside the magma under the crust which causes the movement of the plates. Tell the students that the magma, which is semisolid, heats up when it comes in contact with the outer core and begins to rise. The under surface of the crust is cooler, and when the magma makes contact with it, it cools enough to become more dense and it sinks. This sets up convection currents which act as conveyor belts, carrying the crust along with it.

Closure

Provide students with copies of Convection Currents Move the Crust (page 66) and have them color code the drawing and place it in their geology journals. The colors may be as follows:

*   ✓ Red—arrows in the mantle which are deep and thus the hottest.

*   ✓ Blue—arrows in the mantle that are near the crust and therefore cooler.

*   ✓ Green—arrows on the surface of the crust which are pointing toward each other showing convergent plates.

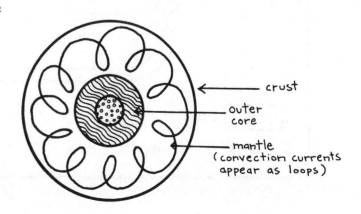

Moving Machine *(cont.)*

Convection Currents Move the Crust

This drawing depicts the convection currents which scientists think take place in the mantle of the earth, just below the crust. These currents are created when magma comes in contact with the hot outer core and heats up, thus becoming less dense. This heated magma rises to the under surface of the crust where it cools and sinks again to the outer core. As this cycle slowly continues, the crustal plates are pushed around like ice floes on water. Some of the plates sink below others and are melted into magma. Magma can also push through cracks in the crust to create volcanoes or ridges.

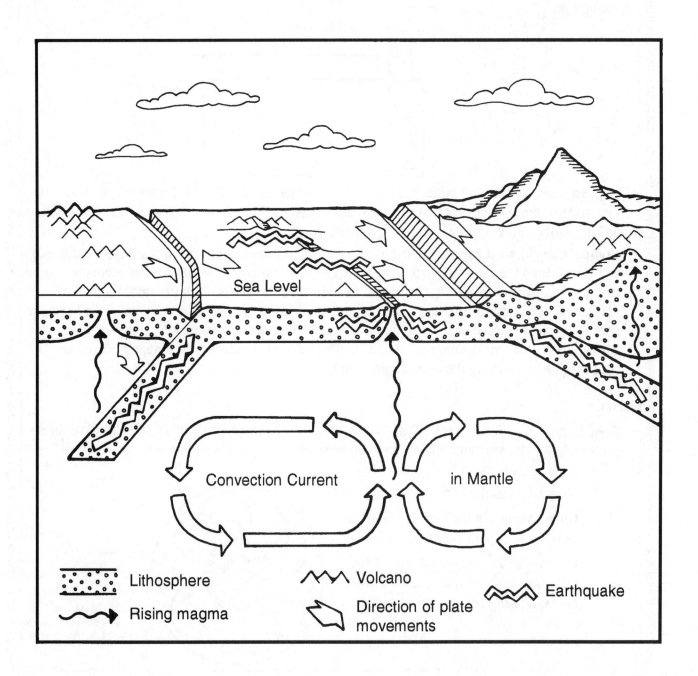

Moving Machine *(cont.)*

Draw your observations from your teacher's demonstration.

Language Arts

Reading, writing, listening, and speaking experiences blend easily with the teaching and reinforcement of science concepts. Science can be a focal point as you guide your students through poems and stories, stimulating writing assignments, and dramatic oral presentations. If carefully chosen, language arts material can serve as a springboard to geology lessons, the lesson itself, or an entertaining review.

One idea is having students create stories about real or imagined events. A list of possible "story starters" is provided on pages 68-69 or you may have students create their own.

Science Concept:

- All was still that night as I lay asleep. Suddenly, without warning my bed began to rock gently back and forth as if it were a rolling sea. I thought it was a dream. When I realized I was actually awake, I ...

Science Concept:

- We were flying over beautiful snow-topped Mount St. Helens. It was May 18, 1980. My family and I were on our way to Washington for a vacation. Suddenly, the top of the mountain exploded with such force that rock and dust shot up into the air as high as our plane. Our pilot put the plane in a sharp turn to avoid this ominous cloud ...

68

Language Arts *(cont.)*

Science Concept:

- A vacation in Hawaii! My dream was about to come true. We packed our warm weather clothes, plenty of swim gear, and took the flight across the Pacific Ocean. It was night as we approached the islands. There was a strange red glow above the big island of Hawaii. I wondered aloud, could that glow be ...

Science Concept:

- I am a scientist aboard a deep submersible vessel moving along the floor of the ocean. We are at a depth of about four miles, where sunlight cannot reach us. The only light is from the spotlight on our vessel. I look through a small porthole and suddenly I see ...

- Encourage students to add illustrations to their stories. When they are finished, have them share their stories with the class. Have students create covers for their stories and place them in the school library or arrange for the authors to read their stories to students in other classrooms.

Social Studies

Geology has played a significant role throughout history. Civilizations have thrived or declined because of changes in and on the earth.

As you guide your students through lessons in history, geography, cultural awareness, or other areas of social studies, keep in mind the role geology has and continues to play. You will find it easy to incorporate the teaching and reinforcement of science concepts in your lessons.

Map skills are necessary for seismologists to locate earthquake sites. Students can develop their map skills while plotting the history of major earthquakes.

Science Concept: Earthquakes occur along the edges of plates in the earth's crust.

Materials Needed

- a copy of the list of major earthquakes from *The World Almanac*

Procedure

1. Divide the information available among groups of students.
2. Have students plot this data on a large world map.

Extensions

- Have students gather information from newspapers and magazines regarding recent earthquakes.
- Have students check other sources for information about historical earthquakes such as:
    - ✓ Lisbon, Portugal–1531
    - ✓ San Francisco–1906
    - ✓ Alaska–1964
    - ✓ San Fernando Valley–1971
    - ✓ Mexico City–1985
- Have students compare the damage in these areas.
- Have students research modern construction methods of designing earthquake buildings and bridges.

Closure

Have students compare the Cracked Earth Map (page 62), with this map to see how many of the sites are near edges of the crustal plates. Then have students compare this data with the map used to plot World Earthquake Epicenters in the activity Earthquakes Around the World (page 43).

Physical Education

What can be more fun for students than being active during a geology lesson. Here is an opportunity to let your students develop their knowledge of geology in a physical way.

Science Concept: Earthquakes are a cause of movement between continental plates.

Divide students into two groups. Have them face each other in two separate lines, pretending to be trees. Explain to students that one line represents the Pacific plate and the other line represents the North American plate. You will represent a large rock between the two plates on a fault line. Tell students when you count to three there is going to be an earthquake and each line will take three steps to their left. While this is happening, you will be spinning around, pretending to be ground into smaller rocks. Explain to students that the land to the west of the fault is slowly moving north and the land to the east of the fault is slowly moving south.

Mathematics

The study of geology requires the use of math skills. Measuring, comparing, and graphing are just a few of the skills that can bring mathematics into your geology lessons.

- Have students practice reading and making charts and graphs.
- Provide opportunities for students to record data on a variety of graphs and charts. Teach the skills necessary for success.

The study of earthquakes provides an opportunity to develop skills in graphing data. Have students use the California Earthquake Data (page 73), which shows the location, date, and magnitude of examples of earthquakes between 1906 to 1992.

Science Concepts: Earthquake locations can lead to predictions of where they are likely to occur, but determining when they will occur is not possible at the present time.

- Have students graph the California Earthquake Data using the date and magnitude of the quake.
- Tell students the graph should be a bar graph with coordinates identified as follows:

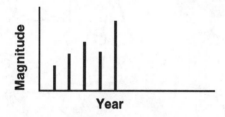

- Divide the data among groups of students as follows:

Group 1:	San Francisco, 1906–Santa Barbara, 1925	19 years
Group 2:	Monterey Bay, 1926–Calexico, 1934	8 years
Group 3:	San Francisco, 1938–Desert Hot Springs, 1948	10 years
Group 4:	Twenty-nine Palms, 1949–Bakersfield, 1952	3 years
Group 5:	Eureka, 1954–Loyalton, 1959	5 years
Group 6:	West of Arcata, 1960–Santa Rosa, 1969	9 years
Group 7:	Borrego Springs, 1969–Santa Barbara, 1978	9 years
Group 8:	Gilroy, 1979–Shandon, 1992	13 years

- Have students use large sheets of .4" (1 cm) graph paper (page 74). The magnitude ranges from 5.5 to 8.1. Stretch this out on a graph by marking the magnitude in tenths (e.g., 5.5, 5.6, 5.7, etc.). This will require 12" (30 cm).
- The years range from 1906 to 1992. Separate each year by .4" (1 cm). This will require at most 8" (19 cm). The bars for the data will need to be thin to permit more than one set of data being plotted for some years.
- Trace over these lines with a black marker so they can be seen from a distance.
- Have students display the graphs in order on the board so they make a continuous graph.
- Tell students to look closely at them to see if they detect any pattern in the frequency of the earthquakes.

Students will find upon close examination that there is no pattern. This is the very problem seismologists are faced with when predicting earthquakes. They do not occur on a regular basis, nor are they related to any astronomical cycle of the motion of moon, planets, or appearance of sunspots.

Mathematics *(cont.)*

California Earthquake Data
(Magnitude 5.5+)
(1906–1992)

Graph this sample of earthquake data to see if there are any patterns which indicate if they occur in a cycle. After the graph is finished, look carefully to see if there are any years when these earthquakes happened more frequently than other years. If so, were the years of greater frequency evenly spread out between 1906 and 1992? They would need to be evenly divided to be considered a predictable pattern.

Epicenter	Year	Magnitude	Epicenter	Year	Magnitude
San Francisco	1906	8.3	Off Cape Mendocino	1954	6.2
Corona	1910	6.5	Maricopa	1954	5.9
Coyote	1911	6.6	Palm Springs	1954	6.2
San Jacinto	1918	6.8	East of San Jose	1955	5.8
San Bernardino	1923	6.2	West of Ferndale	1956	6.0
Santa Barbara	1925	6.2	Loyalton	1959	5.6
Monterey Bay	1926	6.1	West of Arcata	1960	5.7
Bishop	1927	6.0	Hollister	1961	5.6
Eureka	1932	6.4	Crescent City	1962	5.6
Long Beach	1933	6.3	Ferndale	1967	5.8
Parkfield	1934	6.0	Hayward	1968	7.5
Calexico	1934	6.5	Ocotillo Wells	1968	6.5
San Francisco	1938	7.9	Santa Rosa	1969	5.7
Chico	1940	6.0	Borrego Springs	1969	5.9
El Centro	1940	7.1	San Fernando	1971	6.4
Santa Barbara	1941	5.9	Oxnard	1973	5.9
East of Barstow	1947	6.4	Fortuna	1975	5.7
Desert Hot Springs	1948	6.5	Oroville	1975	5.8
Twenty-nine Palms	1949	5.9	Bishop	1978	5.8
Lassen Park	1950	5.5	Santa Barbara	1978	5.7
Off Cape Mendocino	1951	6.0	Gilroy	1979	5.9
San Clemente Island	1951	6.0	Calexico	1979	6.6
Calipatria	1951	5.6	Livermore	1980	5.5
Bakersfield	1952	7.7	Mammoth Lake	1980	6.1
Bakersfield	1952	5.8	Coalinga	1983	6.5
Eureka	1954	6.6	Shandon	1992	6.5

Mathematics *(cont.)*

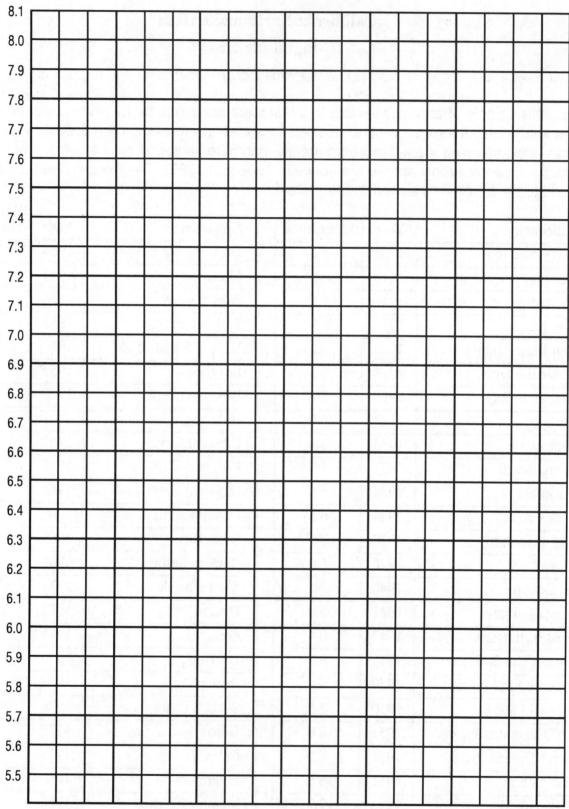

MAGNITUDE

8.1
8.0
7.9
7.8
7.7
7.6
7.5
7.4
7.3
7.2
7.1
7.0
6.9
6.8
6.7
6.6
6.5
6.4
6.3
6.2
6.1
6.0
5.9
5.8
5.7
5.6
5.5

1906 1910 1915 1920 1925 1930 1935 1940 1945 1950 1955 1960 1965 1970 1975 1980 1985 1990 1995

YEAR

Creative Arts

Creative arts offers many opportunities to enhance science learning. Two possibilities are described below.

Science Concept: As the continents continue to move, climate changes will result in different environments which will require changes in living things for them to survive.

The concept is well illustrated and scientifically explained in *After Man, A Zoology of the Future* by Dougal Dixon (St. Martins Press, 1981). See bibliography, pages 95-96.

This imaginative book projects animal evolution into the future, 50 million years from now. Humans have become extinct, and animals presently found on the land, sea, and air will have evolved into very different forms. The continents will be relocated, changing climate conditions and thus habitats to which animals need to adapt for survival. Among these animals are the *rabbuck* (combination of rabbit and deer), *desert leaper* (a huge kangaroo rat), and the terrifying *night stalker*. These are shown in beautiful color drawings complete with explanations of animals' habitats and the adaptations they have made which enable them to live there.

Using the illustrations from this book to inspire students, have them make drawings of other current animals and plants which may evolve 50 million years from now. Let them explain what their environment would be like and how this influences physical changes.

"The Zebirdielioness"   by: Tina B.

Creative Arts *(cont.)*

Science Concept: The earth's crust consists of three types of rocks which are constantly being recycled. The interior of the earth is divided into three layers, growing increasingly hotter towards the center.

Combine art and drama by having students bring to life *The Magic School Bus Inside the Earth* by Joanna Cole (Scholastic Books, 1987). This delightful and scientifically accurate book takes the students of Ms. Frizzle's class on a journey into the center of the earth. Along the way students learn about minerals, rock layers, recycling of rocks, and the parts of the interior of the earth. This book can easily be the basis for a puppet show. Students can create puppets drawn from characters in the book. These can be cut out and glued to cardboard and then fastened to a stick, such as a tongue depressor or craft stick. Students can make scenery for the puppet theater to enhance the show.

Make a simple puppet theater from a large appliance box, such as one from a refrigerator. Tack scenery to the back of the box behind the opening and change it between scenes. This will help students develop their communication skills as well as art skills and should draw upon the talents of all students.

Final Assessment

The activities in this section are designed to assess student understanding of the concepts and lessons taught in this unit as well as the Scientific-Process Skills of:

Applying	Categorizing
Communicating	Comparing
Inferring	Observing
Ordering	Relating

Students will use some of the same activities they used in this book during the assessment. These should be posted in the room or made available at work areas for the students to use. Listed below are questions which require materials:

Question #	Materials
1	Earth's Time Line Information (pages 17–18)
2	Continents Today (page 27)
	Continents: 50 Million Years in the Future (page 28)
	Continents: 200 Million Years Ago (page 23)
	Continents: 65 Million Years Ago (page 26)
4	California and World Maps (page 42) with earthquake locations plotted
	Cracked Earth Map (page 62) showing plates
5	Seismogram (page 47)

Instructions for the Assessment

(Read the following aloud to the students.)

Today, you will do a variety of activities which are related to those we have just finished in our study of geology. You will use some of the materials we used in our lessons. These have been posted around the room. (Show their location.) As you work on the assessment questions, you should not talk or share your answers. I want to see how much each of you has learned from the work we have been doing.

You will not need to do the questions in order. If there are too many people using the materials, go on to a question that does not require the use of maps or other items. If you do not understand a question, ask me for help, not another student.

Do you have any questions before we begin?

To the teacher: You may set a time limit on the assessment, but students should be given enough time so that most will finish.

Final Assessment *(cont.)*

Name: _____ Date: _____

1. Use the Earth's Time Line Information to find the answers to the following questions:

   a. Dinosaurs became extinct about 65 million years ago, but there was an even earlier time when nearly all plants and animals became extinct. When did this take place?

   _____

   b. If you were a visitor from another planet and landed on earth 3.5 billion years ago, what living things would you have found? Where were they living?

   _____

2. Compare the maps of Continents Today and Continents 50 Million Years in the Future. List at least four differences you found:

   _____   _____

   _____   _____

   Compare the maps of Continents 200 Million Years Ago and Continents 65 Million Years Ago. Describe at least four changes which took place during that time:

   _____   _____

   _____   _____

3. Draw a picture of the earth inside the circle below. Label the parts of the earth and tell something about each of these parts. You choose whether or not to color your drawing.

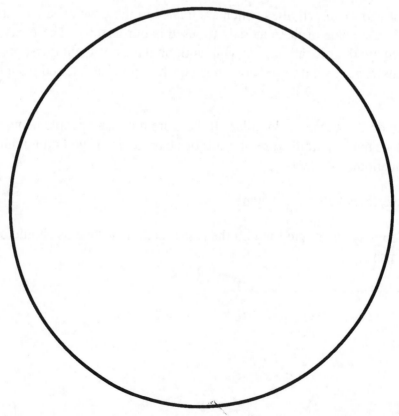

Final Assessment *(cont.)*

4. Look at the California and World Maps on which your class plotted earthquakes and a copy of the Cracked Earth Map to help you answer the following questions:

   a. Why are there so many earthquakes in northern India?

   _____

   b. Why does California have so many earthquakes between San Diego and San Francisco?

   _____

5. Look at the seismogram and describe how it is made. (You may include a drawing but be sure to label all the parts of your drawing.)

   _____

   _____

6. Describe how the earth is like a hard boiled egg. You may use a drawing to show your description but be sure to label all the parts of your drawing.

   _____

   _____

7. Make a drawing explaining how the crust of the earth can move. Label all parts of the drawing.

Science Safety

Discuss the necessity for science safety rules. Reinforce the rules on this page or adapt them to meet the needs of your classroom. You may wish to reproduce the rules for each student or post them in the classroom.

1. Begin science activities only after all directions have been given.

2. Never put anything in your mouth unless it is required by the science experience.

3. Always wear safety goggles when participating in any lab experience.

4. Dispose of waste and recyclables in proper containers.

5. Follow classroom rules of behavior while participating in science experiences.

6. Review your basic class safety rules every time you conduct a science experience.

You can still have fun and be safe at the same time!

Geology Journal

Geology journals are an effective way to integrate science and language arts. Students are to record their observations, thoughts, and questions about past science experiences in a journal to be kept in the science area. The observations may be recorded in sentences or sketches which keep track of changes both in the science item or in the thoughts and discussions of the students.

Geology Journal entries can be completed as a team effort or an individual activity. Be sure to model the making and recording of observations several times when introducing the journals to the science area.

Use the student recordings in the Geology Journal as a focus for class science discussions. You should lead these discussions and guide students with probing questions, but it is usually not necessary for you to give any explanation. Students come to accurate conclusions as a result of classmates' comments and your questioning. Geology Journals can also become part of the students' portfolios and overall assessment program. Journals are a valuable assessment tool for parent and student conferences as well.

How To Make a Geology Journal

1. Cut two pieces of 8.5" x 11" (22 cm x 28 cm) construction paper to create a cover. Reproduce page 82 and glue it to the front cover of the journal. Allow students to draw geology pictures in the box on the cover.
2. Insert several Geology Journal pages. (See page 83.)
3. Staple together and cover stapled edge with book tape.

81

My Geology Journal

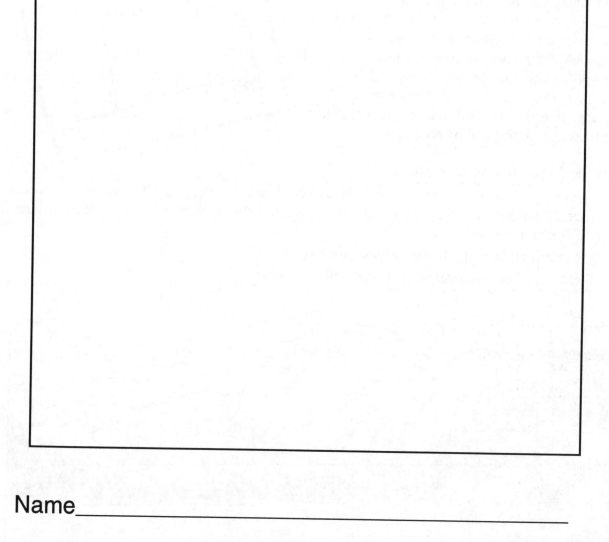

Name_____

Geology Journal

Illustration

This is what happened:

_____

_____

_____

_____

This is what I learned:

_____

_____

_____

_____

My Science Activity

K-W-L Strategy

Answer each question about the topic you have chosen.

Topic: _____

K - What I Already **Know:** _____

_____

_____

_____

_____

W - What I **Want to Find Out:** _____

_____

_____

_____

_____

L - What I **Learned after Doing the Activity:** _____

_____

_____

_____

_____

Investigation Planner *(Option 1)*

Observation

_____

_____

Question

_____

_____

Hypothesis

_____

_____

Procedure

Materials Needed:

_____

_____

_____

Step-by-Step Directions: (Number each step!)

_____

_____

_____

_____

_____

Investigation Planner *(Option 2)*

Science Experience Form

Scientist _____

Title of Activity _____

Observation: What caused us to ask the question?

_____

_____

_____

Question: What do we want to find out?

_____

_____

_____

Hypothesis: What do we think we will find out?

_____

_____

_____

Procedure: How will we find out? (List step-by-step.)

1. _____

2. _____

3. _____

4. _____

Results: What actually happened?

_____

_____

_____

Conclusions: What did we learn?

_____

_____

_____

Geology Observation Area

In addition to station-to-station activities, students should be given other opportunities for real-life science experiences. For example, earthquake models and mineral samples can provide vehicles for discovery learning if students are given time and space to observe them.

Set up a geology observation area in your classroom. As children visit this area during open work time, expect to hear stimulating conversations and questions among them. Encourage their curiosity but respect their independence!

Observation Table

Books with facts pertinent to the subject, item, or process being observed should be provided for students who are ready to research more sophisticated information.

Sometimes it is very stimulating to set up a science experience or add something interesting to the Geology Observation Area without a comment from you at all! If the experiment or materials in the observation area should not be disturbed, reinforce with students the need to observe without touching or picking up.

Assessment Form

The following chart can be used by the teacher to rate cooperative learning groups in a variety of settings.

Science Groups Evaluation Sheet

Room: _____   Date: _____

Activity: _____

Everyone

	Group									
	1	2	3	4	5	6	7	8	9	10
. . . gets started.										
. . . participates.										
. . . knows jobs.										
. . . solves group problems.										
. . . cooperates.										
. . . keeps noise down.										
. . . encourages others.										

Teacher comment

_____

Bragging rights for the group session: _____

_____

Assessment Form *(cont.)*

The evaluation form below provides student groups with the opportunity to evaluate the group's overall success.

Cooperative Group Evaluation

Assignment: _____

Date: _____

Scientists	Jobs
_____	_____
_____	_____
_____	_____
_____	_____

As a group, decide which face you should fill in and complete the remaining sentences.

1. We finished our assignment on time, and we did a good job.

2. We encouraged each other, and we cooperated with each other.

3. We did best at _____

_____ .

4. Next time we could improve at _____

_____ .

Super Geology Award

This is to certify that

_____

Name

made a science discovery.

Congratulations!

_____

Teacher

_____

Date

Super Geologist!

Science Collectibles

Spontaneity is an important element of the hands-on learning experience. As students explore new ideas, they want to add hands-on activities to their science experiences. For this reason it is important to keep a readily available supply of equipment in the classroom, including items most commonly used by students.

Many schools have science kits available for teacher/student use. The problem is that no one ever seems to know where the materials and equipment are! Make use of these valuable resources and supplies by allocating a central closet or classroom that can be used as a science kit storage room. To set it up, have teachers gather school-owned materials and equipment to place in the closet/room. Label each item and provide a system for check-out and return of kit items.

Listed below are items that are often needed but not always provided in science kits. In many cases budget restrictions may also limit the science materials available. Use the supply letter (page 92) to send notes home asking parents for specific items needed for personal classroom experiences or for the science kit itself.

Containers	**Equipment**	**Miscellaneous Materials**
coffee cans	magnifying lenses	aluminum foil
fish tanks/bowls	sponges	rocks
measuring cups/spoons	candles	sand
resealable plastic bags	hot plates	string
buckets, plastic/metal	scissors	fishing line
jars with lids	hand lenses	straws
tin cans (any size)	tweezers	thread
bottles, glass/plastic	measuring sticks/tapes	toothpicks
cups, plastic/foam	timers	(anything else you can think of)
	stopwatches	
	shovels	
	scales	

Supply Letter Request

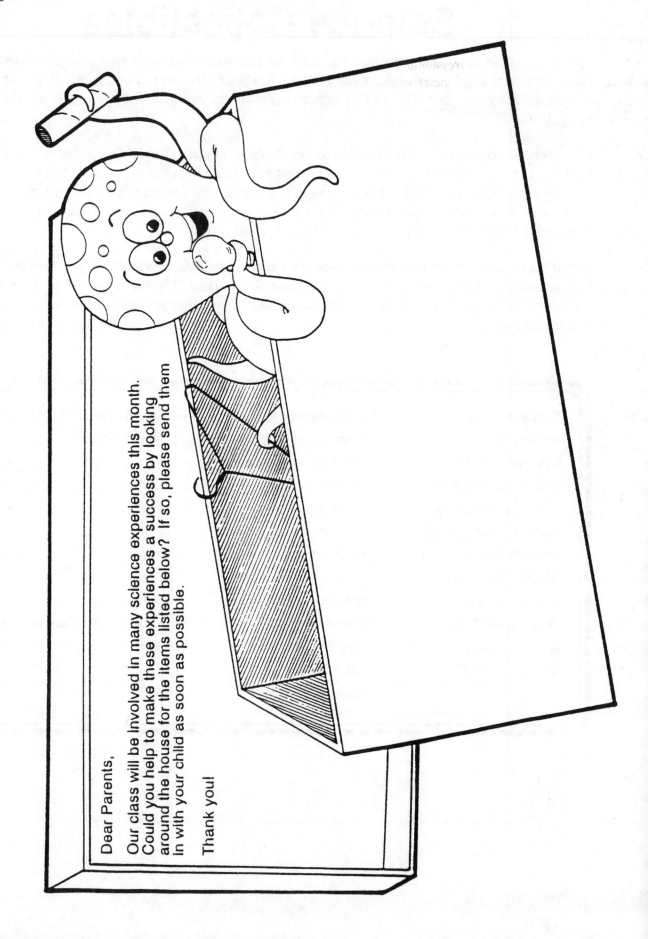

Dear Parents,

Our class will be involved in many science experiences this month. Could you help to make these experiences a success by looking around the house for the items listed below? If so, please send them in with your child as soon as possible.

Thank you!

92

Glossary

C

Compressional Waves—the fastest of the seismic waves which reach the seismographs first and are therefore called primary (P) waves.

Conclusion—the outcome of an investigation.

Convection Currents—the movement of heated gases or liquids between regions of unequal density.

Crust—the hard layer of rock which covers the outside of the earth. This layer is about 5–20 miles (8–32 km) thick and lies just above the mantle.

E

Epicenter—located above the focus on the crust of the earth. Although this is usually the site of the greatest destruction, locating the exact spot of the epicenter requires information from three or more seismographs.

Experiment—a means of proving or disproving a hypothesis.

F

Faults—sections of the crustal plates which slip up, down, or sideways relative to each other during an earthquake. This may take place below the surface of the crust and not be visible from above. These are hidden faults.

Focus—origin of an earthquake, which may be miles below the crust.

H

Hypothesis (hi-POTH-e-sis)—an educated guess to a question you are trying to answer.

I

Igneous Rock—rock formed from melted rock which was originally magma.

Inner Core—the innermost core of the earth, consisting of the heaviest elements of iron and nickel which came to rest here during the early stages of earth's formation.

Investigation—an observation of something followed by a systematic inquiry to examine what was originally observed.

L

Lava—magma which is forced up through cracks in the earth's crust and pours forth from volcanoes.

M

Mantle—the layer just below the crust which is about 1,800 miles (2,880 km) thick and consists of semisolid rock which is approximately 1,600° F (871° C). This extremely hot taffy-like layer moves slowly in convection currents between the crust and outer core.

Magma—the semisolid material in the mantle. When it pours through cracks in the crust called volcanoes, it is called lava. When it cools and hardens, it is called igneous rock.

Magnitude—the intensity of an earthquake, usually measured on the Richter Scale.

Metamorphic Rock—rock which was once igneous, sedimentary, or metamorphic rock which is physically and sometimes chemically changed due to heat and pressure.

O

Observation—careful notice or examination of something.

Glossary *(cont.)*

Outer Core—just below the mantle, this 1,400 mile (2,240 km) layer is made of iron and nickel which is 4,000° F (2,204° C).

Pangaea—name of the land mass of continents that were grouped together 200 million years ago.

Plate Tectonics—the theory which explains the movement of the continents and spreading of the sea floor. The theory is that the earth's crust is divided into many huge plates which are pushed around by the semisolid magma below it.

Procedure—the series of steps carried out in an experiment.

Question—a formal way of inquiring about a particular topic.

Results—the data collected after an experiment.

Richter Scale—developed by seismologist Charles Richter in 1935 to measure the ground motion caused by an earthquake. Every increase in number actually indicates the energy is 32 times greater than the one before it.

Rock Cycle—rocks are constantly being recycled through erosion from water, ice, wind, and heat. All the rocks which make up the earth have been here since it originated about 4.5 billion years ago, except a few additions from meteors which were pulled to the earth as they orbited near the planet. (Also, a few more were brought to the earth by astronauts.)

Scientific Method—a creative and systematic process of proving or disproving a given question, following an observation. Observation, question, hypothesis, procedure, results, conclusion, and future investigations comprise the scientific method.

Scientific-Process Skills—the skills needed to think critically. Process skills include: observing, communicating, comparing, ordering, categorizing, relating, inferring, and applying.

Seismic Waves—vibrations given off by the violent breaking of rock during an earthquake. These waves move out from the focus in all directions, like sound waves. They travel through the body and surface of the earth.

Seismogram—record made by a seismograph showing the time and magnitude of seismic waves.

Seismograph—an instrument used to detect the seismic waves of an earthquake.

Seismologist—scientist who studies earthquakes.

Shear Waves—slower seismic waves which reach seismographs after the P waves. These are referred to as secondary (S) waves.

Surface Waves—a seismic wave that slowly travels along the crust of the earth in a rocking motion before reaching seismographs. These are referred to as long (L) waves.

Variable—the changing factor of an experiment.

Bibliography

Amato, Carol. *Earth.* Smithmark, 1992.

Archer, Jules. *Earthquake!* Macmillian Child Grp., 1991.

Asimov, Isaac. *Earth: Our Home Base.* PLB, 1988.

Booth, Basil. *Earthquakes & Volcanoes.* Macmillian Child Grp., 1992.

Boyer, Robert E. & P.B. Snyder. *Geology Fact Book.* Hubbard Science, 1986.

Bresler, L. *Earth Facts.* EDC, 1987.

Clark, John. *Earthquakes to Volcanoes: Projects with Geology.* Watts, 1992.

   Mining to Minerals: Projects with Geography. Watts, 1992.

Cleeve, Roger. *Earth.* S & S Trade, 1990.

Cole, Joanna. *The Magic School Bus Inside the Earth.* Scholastic Books, 1987.

Cork, B. & M. Bramwell. *Rocks & Fossils.* EDC, 1983.

Deery, Ruth. *Earthquakes & Volcanoes.* Good Apple, 1985.

Dixon, Dougal. *After Man, A Zoology of the Future.* St. Martins Press, 1981.

Fejer, Eva and Fitzimons, Cecilia. *Rocks and Minerals.* Longmeadow Press, 1988.

Gregory, K. *Earthquake at Dawn.* HarBraceJ., 1992.

Horenstein, Sidney. *Rocks Tell Stories.* Millbrook Pr., 1993.

Jennings, Terry J. *Rocks.* Garrett Ed Corp., 1991.

Lampton, Christopher. *Earthquake.* Millbrook Pr., 1991.

Leutscher, Alfred. *Dinosaurs and Other Ancient Reptiles and Mammals.* Grosset and Dunlap, 1971.

Lye, Keith. *Rocks, Minerals, & Fossils.* Silver Burdett Pr., 1991.

Marcus, Elizabeth. *Rocks & Minerals.* Troll Assocs., 1983.

Mariner, Tom. *Earth in Action Series.* Marshall Cavendish, 1990.

McNulty, Faith. *How to Dig a Hole to the Other Side of the World.* Harper Trophy, 1979.

National Wildlife Federation. *Geology: The Active Earth.* National Wildlife, 1991.

Parker, Steve. *Earth & How It Works.* Dorling Kindersley, 1993.

Radlauer, Ed & Ruth Radlauer. *Earthquakes.* Childrens, 1987.

Rhodes, Zim Shafer. *Fossils, A Guide to Prehistoric Life.* Golden Press, 1962.

Schatz, Dennis. *Dinosaurs, A Children's Activity Book.* Pacific Science Center, 1987.

Smith, Bruce. *Geology Projects for Young Scientists.* Watts, 1992.

Stein, Sara. *The Evolution Book.* Workman Publishing, 1986.

Thurber, Walter. *Exploring Earth Science: Geology.* Allyn and Bacon Inc., 1976.

Walker, Jane. *Earthquakes.* Watts, 1992.

Watts & Tyler. *The Earth.* EDC, 1976.

Bibliography *(cont.)*

Spanish Titles

Cole. J. ***El autobus magico en el interior da la tierra*** *(The Magic School Bus Inside the Earth).* Scholastic Book Services, 1987.

Munsch, R. ***Agu, agu, agu*** *(Murmel, Murmel, Murmel).* Firefly Books, 1989.

Natural History Museum. ***Rocas y minerales*** *(Rocks and Minerals).* Santillana Pub. Co., 1988.

Tayler, P. ***Los fosiles*** *(Fossil).* Santillana Pub. Co., 1990.

Technology

Aquarius. ***The Earth Moves, Folds and Faults***, and ***Earthquakes***. Available from CDL Software Shop, (800)637-0047. software

D.C. Heath. ***Dating and Geologic Time, Mountains and Crustal Movement***, and ***The Changing Earth***. Available from William K. Bradford Publishing Co., (800)421-2009. software

National Geographic Series. ***STV: Restless Earth***. Available from VideoDiscovery (800)548-3472. video disc

Troll. ***SimEarth***. Available from Troll, (800)526-5289. software

WDCN, Nashville, TN. ***The Earth, Rocks, Fossils, and Minerals***, and ***Rocks to Rings***. Available from AIT, (800)457-4509. video

Resources

California Divisions of Mines and Geology, Publications Sales, P.O. Box 2980, Sacramento, CA 95812, (916)445-5716, #1 Fault Map of California

Carolina Biological Supply Company, 2700 York Rd, Burlington, NC 27215-3398 or P.O. Box 187, Gladstone, OR 97027-0187 (800)334-5551
Supplier of Biology/Science Materials

Delta Education, Inc., P.O. Box 950, Hudson, NH 03051, (800)442-5444
Science equipment especially suited for elementary grade levels.
Request Elementary Science Study (ESS) catalog for mineral specimens.

National Science Teachers Association, 1742 Connecticut Ave., NW, Washington, DC 20009-1171, (202)328-5800
Membership benefits include subscription to one or more NSTA journals of teaching science which may offer many classroom activity ideas. Members also receive NSTA Reports and an extensive annual catalog of science education suppliers.

Ward's Natural Science Establishments, Inc., 5100 W. Henrietta Rd., P.O. Box 92912, Rochester, NY 1492-9012 (800)962-2660
Supplier of materials and equipment for all science areas. Request catalog for geology.